T0250021

Call Center Savvy

How to position your call center for the business challenges of the 21st century.

By Keith Dawson

Editor of the Call Center News Service, www.CallCenterNews.com

Telecom Books
New York

Published by Telecom Books
An imprint of Miller Freeman, Inc.
12 West 21st St., N.Y., N.Y., 10010
www.telecombooks.com

ISBN 1-57820-050-4

For individual orders, and for information on special
discounts for quantity orders, please contact:
Telecom Books
6600 Silacci Way
Gilroy, CA 95020
Tel: 800-LIBRARY or 408-848-3854
Fax: 408-848-5784
Email: telecom@rushorder.com

Distributed to the book trade in the U.S. and Canada by
Publishers Group West,
1700 Fourth St., Berkeley, CA 94710

First Edition, April 1999

Cover design by Regula Hofmann

Transferred to Digital Printing 2007

table of contents

Preface

Part One: Introducing Your Call Center

Part Two: Technology

Part Three: The Internet Cometh

Preface

This is not a book about call center technology, per se. It's not a handbook to operations. And it's not going to tell you what your service level should be.

Instead, think of this book as a roadmap. This book can help you look at your call center in a new way. If you're a manager, chances are you are used to thinking of your center as a very short term proposition. You are constantly besieged by small problems that you have to deal with before they become large problems. You have to cope with overstaffing and understaffing, productivity and performance, people and technology, all at the same time.

What you're probably missing (and what this book aims to address) is the long view. The ways in which the center develops over time, and how that impacts the company's bottom line. The technologies that are on the horizon, and whether you should be thinking about implementing them now. Perhaps most important, the changing relationship between the call center and the rest of the enterprise — this is one curve you don't want to get behind.

Call Center Savvy is a collection of short pieces, each designed to help you think about an issue in a broader way. In Part One, I set out what I think is a pretty good model for how call centers evolve over time. This model can be used to determine where you (and your industry) fit into the call center spectrum. I also hope it will spark some thinking about the way people use their call centers to deal with customers.

From there, we segue into technology, for a look at some of the cutting edge ideas that are making their way into centers, and a look in particular at three of the most important new technologies: speech recognition, customer information systems, and distributed or "virtual" centers.

What would a roadmap be without talk about the Internet? There are so many question marks about the right role of the Internet in the call center, that there really is no right answer. You have to decide based on your technological comfort level, the comfort of your people, and most important, the nature of the relationship you have with your customers.

Finally, after some consideration to the issues you deal with actually running your centers, we jump off and look at how the call center industry is expanding outward from North America. It's having a tremendous impact on economies and jobs all over the world, something that's not often noted here in the US.

The future, of course, is what you make of it. There's much debate about what call centers will look like as we walk into the twenty-first century: will they be agentless? Will they even exist at all? While no one knows the answers, it helps to be armed with as much information as possible. Use this book to figure out what questions to ask, and to see your center in a different light. It's not just a place where calls are handled; it's the eyes and ears of your company, the connection point, the unheralded strategic asset. It needs to be studied, not just managed. It needs to be understood. This book is a start.

For more information about call centers, check out my website and newsletter, the *Call Center News Service*. It's at www.CallCenterNews.com, and I think it's pretty useful.

Keith Dawson

Spring, 1999

Part One

Technology

The CALL CENTER is BORN
when the volume of calls,
coupled with the need
for a smoother workflow process,
DRIVES the company
to respond with a
more formalized
CALL-ANSWERING system.

chapter 1

Looking Forward, Looking Back

The past few years have been incredibly interesting ones for the call center industry. They've been times of watchful waiting, of incremental change, of taking stock. Times in which the call center has been more widely recognized by the rest of the world — whether they call it the service center or the customer care center or the help desk, years from now we'll look back on the late 1990s as the point when corporate planning began looking at call centers as strategic assets in a meaningful way.

These are the days when we've finally gotten the answer to the question 'Will people buy over the Internet?' — and the answer was a resounding yes. Just ask the folks at Amazon.Com, a company which some are describing as the first all-Internet all-call center company. Hold that thought — more on the Internet in a moment.

We also saw the beginnings of a disjunction between the operating philosophies of outsourcers and in-house call centers. Though much of the rest of the corporate world sees the two as interchangeable, they are clearly not, with different priorities and modes of business. In 1997, at least one significant segment — Wall Street — learned the hard way that outsourcers, though powerful and sophisticated, are not an "index" for the call center industry as a whole. One after another, we saw outsourcing companies take hits in their stock prices because (I believe) financial analysts mistakenly assumed that call center growth and outsourcing growth were the same thing. They work in tandem, but, alas, there appears to be an oversupply of outsourcing capacity in North America now, despite a voracious appetite for in-house call center services.

We have seen the promise of computer telephony integration bear fruit for call centers. And improved products that tie together the call center and the back office processing of customer transactions into a seamless whole — something that used to be just a pipe-dream, or an incredibly expensive customization. Years of open specs, standards and developer's kits have finally brought the ACD onto the corporate network. It is now more of an "information appliance" than most would have predicted a few years ago. This is an unstoppable trend, and a welcome one.

Call centers measure calls. But companies measure customers, and sales. To correct this awkwardness, a new software category has blossomed. Customer *information systems*, a welcome blending of help desk software, telemarketing/scripting systems, transaction processing and middleware, burst on the scene from companies like Quintus, IMA, Point Information Systems, Chordiant, and many others.

The View Going Forward

At one point I tried to make the argument that Internet telephony, despite a lot of hype and some dramatically good new technology, was not a good match for the call center. My feeling was that customers would be reluctant to use Internet resources for things that were more easily obtained by calling a person. Now it appears that Internet telephony is like an oncoming train. It will roll through call centers, telcos, businesses of all types. Newsweek reported that Americans spent $1 billion on-line in 1997, and predicted $38 billion for 2002. That's a dramatic curve, just shy of doubling annually. Consumer behavior is changing faster than I predicted. And the technology is maturing (maybe "mature" is too strong a word, let's call it "ripening").

Prediction for the opening years of the new century: an operational paradigm becomes strongly established in the call center industry for handling some types of customer interactions consistently through the Internet. This will happen on the sales side faster than the service side. The lion's share of the on-line transaction volume will still be "voiceless" for the next several years. Companies will begin to target incentives to customers to use the Internet because of the fantastically low cost of processing on-line transactions.

The ACD will help this process by becoming even less of a switch and more of a network server than ever before. Gone are the days when you would buy a box and put it in a dark room. The typical high-end ACD today is sold with a bevy of links, bridges, gateways, connections and add-on software (often from third-party partners of the switch vendor). RFPs are longer than ever before.

The traditional switch is now responsible for managing interactions from a variety of channels, voiced and voiceless, and for tracking the status and result of those interactions, then passing the data out of the call center into the company's IT infrastructure. Tall order. Watch the switch vendors adapt even more in coming years, with the line between switch, server and network getting fuzzier every day.

Another prediction: long distance carriers, seeing the threat that the Internet poses to their call center customer base, will ratchet up their transaction processing offerings. Look for AT&T and Sprint especially to aggressively target medium-sized call centers with network-intelligence services that take orders and connect to fulfillment and other back-office processes.

Finally, a word about the Asian economic mess. The countries that are in trouble are not now top call center locations in Asia — Thailand, Korea, Indonesia, Malaysia. All along, they were *potential* call center hotspots. The Pacific Rim call center stars have been Australia, Japan and Singapore, plus a few others. Some are speculating that the Asian situation is too unstable, and that it may contract consumer demand in so many ways, in so many countries that the entire call center industry is affected, worldwide.

If more bad economic news percolates out of Asia in the next few years, we may finally know just how recession-sensitive the global call center industry is. I think that in worst case for the US segment, call centers put off major capital purchases (ACDs, new centers), and concentrate on optimizing existing ones.

Which could be good news for software developers, CTI companies, middleware, and customer interaction/data mining companies. And in the best case, the industry keeps on growing worldwide, spreading out its efficiencies, virtualizing across geographically disperse call center networks. My guess is that call centers will turn out to be pretty resilient in the face of bad news from Asia (and anywhere else it turns up). Call centers are a great economic engine, as I'll argue elsewhere in this book. They are powerful tools for companies, countries, regions and counties. They eat up technology and spit out jobs, and they provide you with your single greatest tool ever for connecting with customers. No matter how the form of the center changes, the future belongs to the call center.

chapter 2

The Six-Stage Model of Call Center Development

It's hard, when confronting the day to day managerial headaches of running a call center, to stay focused on the long term, or to see the big picture of where a center fits into the whole corporate structure.

The success or failure of a company's customer relationships depends on more than whether that center varies from preset service level targets today, tomorrow and next week. It depends on how well the call center management team coordinates all the technologies available in service to the company's goals.

And it depends on how well the managers articulate to the rest of the company (including upper management, financial analysts, marketers and product development) a strategy for integrating the call center — and the customer — into the company's daily operations.

Over time, we have seen that call centers tend to follow very similar trajectories in developing their approach to customer relations. Not everyone follows the same path, obviously, and when you have as many call centers globally as we do, where there is room for a trend there is also room for many counter-trends. But we have identified a rough six-stage model for the development of customer service through call centers. As you will see, it has less to do with the deployment of particular technologies than it does with the company's changing philosophy of service and customer relations.

1. Startup. In a company, usually a smaller one, that has just released their first product or a major upgrade. The phone is ringing off the hook, and there are few or no people whose job or training is in dealing with customer problems. And yet the phone needs to be answered, so someone is assigned the task of receiving those calls and logging them, but often not dealing with them. This stage is characterized by a somewhat desperate and surprised recognition all calls must somehow, eventually be answered (never mind when).

Typical technology at this stage: a standard business PBX. No integration between the PC application that the CSR is using and the phone system. (It's even premature at this stage to call the person a CSR — he's probably an entry level person of some sort whose unlucky task is to man the phones.)

2. Triage. Someone at the company recognizes that something has to be done. Customers that don't get a response to their queries call their sales reps, who may or may not be equipped to handle post-sale follow-up. The volume of calls, coupled with the need for a smoother workflow process, drives the company to respond with a more formalized call-answering system. The call center is born.

Typical technology: a departmental ACD spun off the PBX, or in some cases a full ACD with rudimentary reporting features. Sales force automation software coordinated with the outbound side for the salespeople, and some version of that data routed to the inbound reps' desks, but not automatically.

At this stage, when you add technology you are doing it reactively — there are problems that can be remedied by a quick application of some very simple technology (a switch, some software, maybe an outbound dialer). It's an easy cost-justification, but only on the cost control side. It isn't until a later stage that the people running the center begin to think beyond cost control and focus on the revenue possibilities.

Another kind of center that typically functions at this level is the small-scale outbound center, the kind used for local collections, telemarketing or surveys. (Note that we exclude outsourcers from all of this step-model. They run on a completely different course.)

3. The Organized Center. The realization sets in, usually as call volume escalates and the costs of labor along with it, that some kind of organized structure needs to be imposed on what had been a reactive, ad hoc function.

The typical response in stage three is to implement what we now think of as "traditional" call center tools. That includes real ACDs for call routing, and some more formalized grouping structure to go along with it. Division of personnel into first-line sales qualifiers and service reps might also happen late in this stage.

Some sort of front end is also often implemented, first as part of the auto attendant system if they are still running their center on a PBX/ACD, then as they gain sophistication an IVR system might make an appearance. In short, this stage sees the center become formal, with its own management structure and some basic call handling automation.

It's at this stage, interestingly, that most of the call center industry's vendors start their sales process, because this is where most centers appear on the radar screens. It's only when a center reaches this point that the manager starts reading about call center technologies and may go to a trade show or seminar to learn more. As the market becomes more competitive on the vendor side, with more attention paid to the small and developing informal centers, vendors may seek a way to identify centers while they are still at stages one and two. 4. Continuous improvement. Triage recedes into the background and the center chugs along for a while, responding to periodic peaks and valleys and identifying the company's baseline service level; this gets factored into the costs of running the center, and a paradox is reached. The center, analyzed from outside by managers in other departments, is asked to do two things at once — improve service, and cut costs.

It is at this point that the call center management hierarchy begins to look outward for other methods of improvement. This can mean searching for more automation technology in an attempt to cut down on call volume, or call duration. Sometimes CTI is contemplated at this stage.

Or, it can mean bringing in training and manpower assessors to determine whether the workforce can be managed better, at a higher level of efficiency (sometimes this is seen as a quick route to solving both ends of the paradox — but quick it ain't).

Also, workforce management software is sometimes seen as a way to smooth wring some efficiency from an unpredictable ebb and flow of volume.

All of these choices are subjected to rigorous cost-justification, and often they don't pass muster with the managers outside the center. At this stage of development, there has not yet developed an understanding of the mutual interdependency between the company and the center.

5. Corporate asset. As the company itself matures, select people within it begin to realize that the relationship between the company and its customers is of paramount importance. Where this realization sets in differs from company to company: sometimes it's inside the call center, sometimes it comes from upper management on down, sometimes elsewhere.

What results, though, is that the company starts to look to the center as a place of opportunity: as a tool to assist other company strategies like cross-selling or up-selling callers on featured products. To a company in this stage, the call center looks less like a way to triage a flood of calls and more like a corporate asset.

Typical technologies at this stage: screen pop, and perhaps other basic CTI applications; industry-specific verticals built for the call center; on the service side, some basic stabs at alternate inputs, like an e-mail or web interaction system. More likely, though, is that the company would be integrating call center systems with existing company-wide data systems, like transaction processing, order fulfillment and tracking, and workflow.

6. Mass customization of service. Ultimately, the company and its call center become a tightly integrated whole; so whole, in fact, that it might be difficult to tell exactly where to draw the line between the two. Front-end data about the customer is captured automatically, processed, and integrated with the data residing in back-end databases, wherever they might live.

Agents have, at the ready, all the information they might need to process that call, and at this stage when we say call we mean voice, fax, e-mail, chat session, or whatever. They also have the power to resolve any questions or difficulty that might come up, without transferring the call to another department. Each call is customized to the precise needs of the customer, and can be measured many ways: in terms of agent performance, cost of the call, or best of all, by the relation between the cost of processing and the value of the customer.

At stage six, we have reached the theoretical end point of call centers as a customer service delivery tool, because beyond that point the center ceases to exist as separate from the company.

As companies move from stage to stage, they may acquire or cast off certain technologies that are common in one stage or another, but it's my sense that adoption of a particular technology is driven more by industry- and market-specific concerns than by passage through these stages.

It's important to note that this six-stage model was not derived from a scientific study or survey of centers; it's just an observational model that will not fit all or even most centers all or most of the time. But it's useful as a way to peg your own center's philosophy of service delivery, and try to improve by moving to the next stage.

In the next chapter we will discuss where some industries tend to clump in this model, and how to move a center further down the line.

chapter 3

Moving Your Center Forward

Often, companies try to improve their call centers by throwing technology at a problem. Sometimes this works, but not all the time.

In our six-stage model of call center development, we identified the path a company travels in moving from a simple reactive mode to a more intelligent, company-wide strategy for dealing with customers.

Here's a recap of how the model runs:

1. Startup. The phone is ringing and someone has to answer it. So someone does.

2. Triage. The phone is ringing more often than there are people to answer the call. People are pulled off other jobs to answer the phone. Or calls go unanswered and customer dissatisfaction grows. This is the tipping point between informal and formal centers.

3. Organized centers. Traditional call center technology is brought to bear to fix the problems in stage two. In comes the first ACD. Agents are organized into groups.

4. Continuous improvement. It is discovered that call centers cost a lot. Bodies in seats consume precious resources. Management makes the now-famous contradictory demands: cut costs; improve service.

5. Corporate asset. The smart call center manager realizes that the flip side of cutting costs is convincing the company that the call center is generating revenue, and figuring out how to quantify that. In comes more advanced technology, CTI and CIS in particular.

6. Mass customization of service. The nirvana-like state where every customer interaction is handled with the precise pieces of information available (from anywhere in the company) that are needed. (More about this in a moment.)

Last chapter we detailed what happens in each stage and what distinguishes one stage from another. One of the interesting things about this model (and that sets it apart from other developmental models that we've seen) is that it doesn't rely on technology. There is no piece of hardware or software that can magically move a call center from stage two to stage three.

Instead it uses the relationship between the call center and the rest of the company as a way to define the progress of the center. Yes, there are particular technologies that can make or break you — the installation of the first ACD is certainly one of them. You will be a different kind of company the day after your ACD goes in from the day before.

But that's taking the short view. If there's one thing the spectacular development of the internet has taught me, it's that companies benefit from a wider distribution of internal information.

The more someone in marketing knows about what's going on in the call center, the better off they are. And the more the call center agent knows about the products she has for sale, the more likely she'll make a successful upsell or cross-sell.

Working the Model

So how does a manager position his or her call center for growth and continued success?

Even allowing for the fact that this model does not describe every call center in every case, there are some things we can learn from it.

We start from the proposition that stage one is nearly a lost cause. There is no way you can remain informal for long before the demands placed on the "center" ratchet up beyond its limited ability to deal with them.

The evolution from one to two is so natural and necessary that it's not really worth describing.

Two to three, however, is one of the most important development points any center will go through. When a center formalizes, it lays down the template that's going to guide its growth through the rest of the series. Decisions are made at this point that affect a whole range of issues: data management, for example, or the kinds of training that agents and supervisors get.

Unfortunately, centers that are graduating from informal to formal often make these decisions on an ad hoc basis rather than as part of a concerted plan.

Later, when the center (and the company) are prepared to deal with more complex issues of advanced customer management and retention, the decisions that were made early on can be a roadblock to success. You may be locked into particular technologies, for example, that prevent you from moving quickly to adapt to changing market conditions.

Or you might have thrown all your resources behind one vendor who is not capable of handling all the needs of a growing call center.

Even worse, the call center may have institutionalized "bad" behavior — strategies for call handling and customer service that may have worked when the company was simply trying to triage calls, but no longer reflect well on the company in changed circumstances.

While the transition from a stage two call center to one in stage three may seem like a natural step, it's important to remember that the decisions made casually at that moment are critical ones that will have impact far down the road.

Using the Center Strategically

The transition from four to five, though equally important, is handled somewhat differently. Call center managers are constantly grappling with that paradoxical dilemma of cutting costs and improving service. And the solutions that they find are intensely dependent on their unique business circumstances.

Some industries are less sensitive than others to service conditions. Let's face it, until recently there was no reason for most utilities or cable companies to put service higher than cost containment. In financial services, where call centers are used for an immense volume of transactions (both informational and monetary), cost containment meant automation, which was seen as a way of putting more power into the customer's hands — to control their accounts through IVR, or trade stocks over the web.

Within any industry there are examples of centers that operate everywhere on the quality spectrum. It's my feeling that the delicate evolution from four to five has almost nothing to do with the kind of technology used in the center, and everything to do with the mindset of the company's management, especially the call center manager.

When the call center is seen as a tool for generating revenue and for helping quantify the value of each customer to the company, then the center can be used strategically to further the company's goals. In every industry and company those goals will differ, and so will the method of reaching them.

Some adopt cutting edge technology like web-commerce systems, computer telephony or the new data-mining/customer information systems. Others achieve similar ends by focusing on the human: techniques for better coaching of reps, or example, or incentives for keeping people on the job longer to reduce training expenses and agent turnover.

Both approaches (and others as varied as the call center industry) are valid signposts that a company's call center is moving beyond its traditional role as a single-function department.

In a company of any size there may exist several centers, each with its own priorities, resources and technologies. In order to really make it into stage five, those centers have to be coordinated and managed as a group.

This does not necessarily mean that they all handle each other's overflow, or that calls bounce around the network waiting for an available agent at any center (although that's a nice scenario, and more common than it used to be).

Instead, it means that the centers, though physically separate, coordinate their response to customers. That a center devoted to service has access to the customer data entered by an agent at the center that sold the customer something.

Stage Six

No call center is really at stage six. Because when you reach this point you're not really a call center anymore. The industry is constantly buzzing with new terms to describe the call center — everything from "commerce center" to "customer contact zone".

What people are grasping for is a way to describe what happens when anyone in a company is equipped to handle any call that comes in, because all the data pertaining to the interaction is right on their desktop.

As data moves around more freely within companies, helped by the beauty of browser frontends, stage six will mean that a "call" (which now includes e-mails, web interactions, live chat sessions, and so on) will reach its most appropriate recipient.

Appropriate used to mean the person who is best equipped to handle the call. That's stage five thinking. Because at stage six, "best equipped" also means that the interaction is analyzed for value before it happens. If a senior VP is the person best equipped to handle a call, that doesn't mean the customer is valuable enough to warrant the SVP's time.

But maybe she does — the analysis will tell how important that customer is and what that customer's history is like. It will have workflow engines behind it that enforce consistent interactions, whether they are handled by a traditional call center rep or someone outside the center.

In stage six, distinctions break down.

The qualities that distinguish the hypothetical stage six call center have almost nothing to do with the technology at use, and everything to do with how the company is using those technologies. Of course, the machinery makes a lot of exciting things possible.

I've never seen, or heard of, a true stage six call center operation. (Any that would like to claim the status should call us, because we'd love to talk about it.) It's only a matter of time before stage six, or something like it, becomes commonplace.

Of course, when it does, we'll have to invent a stage seven.

Part Two

Technology

What should you be thinking
about when BUYING
CALL CENTER TECHNOLOGY?
INTEGRATION–with every
other piece of hardware
and software in your center.
Mostly the software.

chapter 4

CTI &
The Call Center —
The 2% Solution

Say "computer telephony integration" to most managers and the first thing that they'll think of is "screen pop." Unfortunately, that's often the last thing, as well. Despite great strides in the development of technology, and the creation of applications for that technology, there are still far fewer real-life examples of the kind of intricate, top-to-bottom CTI call centers than we thought there would be by now.

Kevin Kerns of Apropos Technology (formerly Teledata Solutions) told me in 1997 that CTI was then in use in 2% of call centers - or less. That figure includes "informal" call centers, like help desks. Other estimates then and since have ranged as high as 10-14%. Still, that's a pittance. Why so few?

The first reason is because of the sheer size of the market. Best estimates are that there are somewhere between 100,000 and 140,000 call centers in North America. Most of those are small- to medium-sized centers, and a large portion are those so-called information centers. At the top of the pyramid, in the range of the very largest — at the banks, reservation centers and service bureaus — sit many of the centers that are already CTI-enabled.

Another, subtler reason has to do with the way the technology is presented to call center managers — the people who have to buy, recommend, and use CTI on a daily basis. CTI technology, in the form of sophisticated features like skills-based routing, network intelligence, and yes, screen pop, have come to market faster than the user community can assimilate. They have simply not yet figured out how to integrate these capabilities into the daily operation of their centers.

That will certainly change as the community becomes better educated. Vendors and systems integrators need to pay more attention to how CTI installs affect both agents and upper managers. One call center manager at a bank recently told me that when his company installed CTI, the biggest mistake they made was not paying enough attention to the "human factor" - the agents who had no idea what was happening, or why it was good for the company.

Upper managers need to be sold on the cost-benefits of the technology. Vendors need to make clear that the savings are real, not theoretical. And that they can help turn a company's call center into more of a strategic asset. The future vision of much of the industry centers on a picture of the call center linked so tightly into the rest of the enterprise that information passes seamlessly between departments; whenever contact with a customer occurs, so much "intelligence" can be brought to bear on that interaction that it will almost always go smoothly. The vision is so tangible, it's easy to forget that linking phones and computers is hard stuff.

"CTI is definitely in the early adopter phase," says Kerns. "It's still about five years away from enterprise deployment." (Again, he said that in 1997.) But the tantalizing possibilities are beginning to see the light of reality. One of Apropos own customers, US Robotics, used CTI to shorten call duration. They've seen their call volume increase by 600%, but only had to increase the number of agents by 350%. Real savings of $6 million a year. Customer wait times were cut from 12 minutes to two.

Numbers like that help illustrate how CTI is one of the few things that can simultaneously cut costs and improve service. Yes, screen pop is at the heart of most apps, because few things speed the call better than a screen full of data. But it also lays the groundwork for the technologies at the next level, the multi-site routers and dynamic call processors that lie ahead.

And it says to the 90-98% of call centers without CTI: now you are playing catch-up.

chapter 5

A Component-Based Architecture

This is a rather esoteric subject for call centers, but it's actually a high-impact one. It's something that goes on under the hood of many of the major software purchases that a call center makes. Building software using a component-based model is a developing trend. It promises to make software less bloated and more feature-focused. It also may lead to faster deployment of upgrades and easier customization.

Components can create the software you need, tailored to your specific call center transaction processing situation, without forcing you to buy from a consultant who will make software just for you.

Component-based architecture is like object development taken to the next level. Componentization allows a developer to break the program apart into (relatively) tiny parts: the interface would be one, the workflow another, and so on back down the line. In a banking application, for example, the application for processing a stop-check order is its own component, and that in turn integrates with the data source and the host interaction systems.

Technically, component-based systems are easier to upgrade and deploy than non-component systems. Ideally, if developers wanted to change a feature, they would have to code and test only the one component that contained that feature. If the process for stopping a check changed, for example, only the stop-check app would need to be fixed. Every other component stays the same - no more extensive compiling, testing, reinstalling the entire system at the customer site.

On top of that, componentization allows for more flexibility at the end-user. Individual users can be equipped with different front-end GUIs - because the GUI is simply a small component of the whole, creating separate front-ends takes far less time than creating different versions of the program. Everything behind the GUI is the same no matter who is using the system.

Componentization also gives developers the ability to create very customized applications that work off of a common system. A customer can come to a developer and ask for something very specific to their business process — and the developer can say yes, and deliver it with a quick turnaround. That's because the rest of the system remains untouched.

Components are created for functional things like order fulfillment, different kinds of data retrieval and display, or creating quotes. It's made possible by Microsoft's Component Object Model (COM) and Distributed Component Object Model (DCOM) and ActiveX technology.

Creating a component-based system is difficult in the first iteration — an existing program needs to be rebuilt from the ground up. For Versatility, one of the first companies to implement a component-based model for building call center software, it took them more than two years to revamp their existing product.

Versatility's software began life as an outbound-based telemarketing management system, including such features as campaign management, and outbound dialing. It's far more advanced now, but what it retains is its customer transaction focus.

Unfortunately for Versatility, componentizing didn't help them as a company; they had troubles and ended up being bought by Oracle for a song. But the model still holds. In an environment like a call where so many different kinds of applications (both horizontal and vertical) can be used, and every transaction type spawns it's own frontend and backend data events, breaking complex software into smaller pieces for faster deployment is almost a no-brainer.

With all the attention paid to call center/enterprise integration, perhaps the move to component architecture will let companies integrate their processes more tightly without losing sight of the importance of the customer interaction.

chapter 6

Sweet Suites Inter-Application Automation

It's becoming increasingly clear that call centers are not using as much CTI as the industry thinks they are. While the call center industry is the biggest consumer of the technology, penetration of CTI is still as low as 10%, give or take a very few percentage points.

That being the case, what's the best way for a call center that's sold on features and productivity benefits to implement? That depends on size, complexity and the age of the existing infrastructure. But more and more, it's likely that the most robust solution comes from a vendor in that fuzzy category called middleware.

These vendors, who specialize in creating systems that float between layers of technology like the oil in your car's engine, are entering a period of maturity. Their products, no longer 1.0 or even 2.0 releases, are more feature-rich. And most important, they are more seamlessly integrated with the hardware on one side, and the database infrastructure on the other.

This is important, because the installation of CTI in a call center is quickly becoming one of those standard-issue horror stories that scares everyone off a good technology. (You know the horror story I mean — the one where the company asks for screen pop, and the systems integrator promises it'll be installed in six weeks; nine months and millions of dollars later, *something's* installed, but nobody knows exactly what...)

Call centers are scared of systems integrators because they know that a custom solution is asking for trouble. Technology has moved so quickly that no company missed out by waiting to install CTI until it was "call center-ready." The smallest centers (those that are informal, and formal ones under 40 seats) have a wider range of call processing options available on the standard phone system, the PBX, than ever before. I've never seen so many separate functions crammed together — IVR, ACD, auto attendant, unified messaging, openings to the intranet — in today's basic business phone systems.

At the higher end, the high-performance switches for call centers are now so open that it's possible to mix and match add-on software for virtually any vertical or horizontal feature preference without sacrificing switch performance. And without waiting for a vendor to craft a custom integration. Recent announcements by US West and Ameritech of enhancements to their partnership programs, in which they each, separately, partner with a Chinese menu of companies from across the technology spectrum to provide turnkey call center systems, bears this out. There is enough confidence at the highest levels to assure an end user customer that the integrations are solid. And there is now the expectation that so-called CTI features-screen pop and skills-based routing especially-will work, from day one, right out of the box.

Whereas in the early days the spotlight shone on the switch vendors who opened their boxes, and then on the standards-makers who enabled the software, nowadays the attention is on the people who make the applications. CTI moves up the food chain. And the higher it gets, the more it interests call center people. In fact, it's only interesting insofar as it does two things:

• Allows call centers to perform tasks that it couldn't before.

• And gives the management of those centers the return on investment to justify the heartache of installation.

For example, take one familiar CTI system. (We use them as an illustrative example, not because they are any more wonderful than their competitors.) AnswerSoft's Concerto family present CTI to the call center in a user-friendly package. The basic software is SixthSense. This is what most people think of when they think of CTI middleware — it's the software that performs the basic data/voice transfer. It creates a pipeline between the phone switch and the back-office databases, through which data about the call is sent for analysis and caller info is sent to the agent's desktop. It sounds simple, but as we have come to learn, creating that pipeline has been a tremendously difficult challenge for the entire industry.

Now that the pipeline has been created, we can change the way we think of it to make it even more powerful. In the days when developers ruled the CTI universe, the link was sold to the industry as a connection between two pieces of technology, the switch and the database. But when you cross the threshold into the call center, what people want to hear about is function: how will this enhance what I am already doing, and allow me to do more?

This pipeline is, in fact, a way of connecting not two pieces of technology, but two diverse business realms: the inside of the call center and the rest of the organization. They speak different languages and work with different kinds of information, but their mission is intricately tied together.

AnswerSoft appears to understand what call centers want to do with that pipeline. Their suite includes software like SmartRoute, a system for determining where to send the call based on data retrieved from somewhere else in the company. And SoftPhone Agent, which puts the call control features on the agent's PC, and coordinates it with recordkeeping about agent status. Remember that these features were once the province of the switch vendors but have now passed very clearly into the realm of middleware.

AnswerSoft's role is to stand between the front office apps and the back-end databases, in coordination with the switch, to integrate all the current systems into one common environment. One component, Shadow, is a management system for "cradle-to-grave" customer and information tracking and statistical reporting. It measures call center performance in real time, captures business data associated with calls and pumps that data out in a form that management can use. They have added Web support to the automation matrix, treating alternate channels of information as merely an extension to the pipeline in and out of the center.

Smartly, AnswerSoft has created a tool that they use in sales presentations that performs an ROI analysis on particular call center situations — you enter the parameters for call center operation and it spits out details on how many seconds per call you'll save, and translates that into hard dollars savings by month or year. And the user can change the underlying model. This gives the call center manager, who we know wants to buy the technology, the cost justification to take to the financial authorities who disburse the cash. This technology is quite expensive. The sample I walked away with showed an outlay of nearly half a million dollars for an AnswerSoft investment. But the sample also detailed a 4-month payback (in a simulated center, not every case), and more than $7 million in savings over three years.

They also have a promotion where they will go into a call center and create an up-and-running pilot of their CTI system in just a week. Both the pilot promotion and the on-demand ROI quotes go to the heart of the problem. Call centers want finished products. They want those products to fulfill a defined business need (and not be technology for its own sake, because it's cool). And they want those products to show demonstrable hard dollar benefits in productivity enhancement and cost savings. Until now CTI has been unable to provide these things. AnswerSoft's latest release points a direction for these "pipeline" vendors, and ratchets up the competition. That, of course, benefits everyone.

Of course, a lot of these things were reasons that AnswerSoft was purchased by Davox, a company that makes primarily hardware and software for outbound dialing. Why is this a good fit? Davox desperately needed to add the CTI glue to its hardware, to make the argument that their call processing engine can thrive in both an inbound and outbound world. They needed the connections not just to other vendors products, but to a whole range of peripherals and add-on applications that AnswerSoft can provide.

This dynamic is replicating itself throughout the CTI world, where the companies that make middleware are finding themselves welcome partners in pulling some of the older, more telephony-oriented companies and products into the data networking and computing world. As call centers become more "central," handling more different types of interactions that are often part voice, part data, owning the middleware glue becomes critical to success.

chapter 7

Forecasting, Simulating & Routing

Workforce scheduling is a great idea for call centers. It's practically a no-brainer — knowledge about staff plus historical data about call volumes equals a prediction about the future. It's a great help to any manager to be warned that next Thursday between 10:00 and 10:15 he's going to be 17 FTEs away from a poor service level.

These products have been around for a long time. Companies like TCS, IEX, Pipkins, Blue Pumpkin and others have all offered various workforce management systems, some with predictive qualities, others focusing more on scheduling.

What surprises many people is that these products have penetrated into no more than around 10% of call centers in the US, if that many.

Workforce management is expensive. It's been a high-end sell from lots of these companies. It's been the kind of thing that could be justified when purchased with a major switch investment, or for a huge center that makes real money off the staffing savings it brings. But for the broad, 50-seat and under market, this taken-for-granted technology has been largely absent.

It's about more than just price, though. Lots of technologies are expensive. CTI, the Internet, it's all expensive. History shows that if you can demonstrate a clear return on investment within a reasonable period, you can justify buying the technology. Workforce management can definitely show short payback periods (as short as 90 days, in some cases).

One reason for the reluctance (beyond the fact that paying $100,000 for this seems excessive to a lot of people, no matter how productive it'll make them) is that it copes poorly with a skills-based routing system. Recent enhancements to some of the existing products have made some inroads, but none of those newcomers has been battle tested by enough centers to tell us whether they are good solutions to a longstanding problem: true forecasting and scheduling based on agents skill sets.

Back to Basics

It would help to identify exactly what we expect from a workforce management system:

1. It should read historical information from the switch or the network and project the future: what will my traffic and volume be like under similar circumstances?

2. It should hold information about staff — who they are, when they like to work, what group they belong to.

3. It should coordinate those two piles of data and allow a manager to create a periodic schedule, in less time than it would take the manager to make the schedule on a whiteboard, or in Excel.

It should go without saying, but it doesn't, that that the schedule it creates should be a better one than what you can do manually. Better, defined as more precisely matching the agent pool to the call volume so as to minimize staff cost without negatively impacting service level.

4. It should tell you what the results of that schedule were in real life. This can be in the form of real time readout of what's going on in short-term increments (the last fifteen minutes, for example). Or it can be expressed through adherence reports.

5. It should allow the manager to override certain aspects of the schedule as he or she sees fit, without trashing the whole thing.

6. It should scale up to more complicated scenarios: agents spread out among multiple centers, for example.

7. And finally, it should reflect the reality of modern call center life, that agents are often selected for a call based on criteria other than simple availability.

Most of the products on the market are right there for criteria one through six. Number seven, though, is a tough nut to crack. Forecasting based on skill set, or really on any criteria other than random availability, is not possible using the mathematical model that underlies workforce management software.

Unfortunately, skills-based routing is wildly popular and is a feature packaged with every ACD on the market. The question then becomes, if you can't forecast based on historical call volumes, how do you project the future accurately enough to create a schedule that works?

Ofer Matan, of Blue Pumpkin, explained the problem to me this way: Given that we know the values for things like call volume, talk time and the number of agents, we can predict the service level, and then adjust the prediction to get what we want. But in a case where each agent belongs to more than one group, the problem becomes "high-dimensional" — that is, there is more than one solution to any problem.

It then becomes impossible to use an existing algorithm to calculate the optimum parameters, and hence, the best schedule.

Simulating An Alternative

One promising alternative has been simulation. Systems Modeling, a company deeply involved in simulation technology for call centers and other industries, uses this definition:

"Creating a computer model of a real or proposed system and conducting experiments on the model to describe observed behavior and/or predict future behavior before investing any time or money."

Instead of using history as a guide, you create a working model of a situation, in this case a call center. You guess at the values for a series of parameters, and set the simulation in motion to see how it plays out. Sometimes you'll be right, other times wrong.

To minimize the chances that you'll be wrong, what workforce software has to do is allow the user to load the front end with all the parameters that are acceptable and then insulate that manager from the horrible process of scenario refinement. (Remember, this software has to be easier to use than a whiteboard.)

But hidden within the bowels of the software has to be a technique to continually refine the simulation until it comes as close to the target service level/cost parameters as possible.

We are headed in this direction. For several years now, companies like Systems Modeling, Bard Technologies and others have been working on simulation models for call centers. These models have always been very good at helping managers play what-if games: how will adding three trunks affect service level? What if I implement IVR as a front end?

Some of the workforce management systems have turned to simulators to augment their existing prediction models.

This is a category that for years was crying out for some new products, and frankly, for a more customer-centric approach to the problem. Now we are starting to see some sharp products from vendors that are relatively new to the call center industry. And simulation is finally flowering as an alternative to the historical number crunching for prediction.

chapter 8

Predictive Dialers Roll On

Predicting the end, or even the decline, of the high-volume automated dialing system is a mistake. No matter how technology changes, the dialer remains part of the call center landscape for two key reasons:

One, there is a demonstable need for the raw horsepower it can project in an outbound campaign. And two, it's proved surprisingly adaptable to new modes of organizing call centers.

Today's dialers are sophisticated call-crunching powerhouses. They have more in common with the incoming ACD call routers than they do with the closed, proprietary hardware of the early '90s.

In an outbound calling campaign, they screen out the busy signals, no answers, answering machines and Standard Information Tones (SIT).

In fact, predictive dialing automates the entire outdialing process, with the computer choosing the person to be called and dialing the number. The call is only passed to the agent when a live human answers.

(This is a higher-tech alternative to what's known as "preview" dialing, or screen dialing, where the rep sees the customer information and then, when ready, places the call. There are other, hybrid dialing types, but predictive means something very specific, and vendors shouldn't use that term unless they mean it.)

What It's Good For

Predictive dialers screen out all the non-productive calls before they reach the agent: all the busy signals, no-answers, answering machines, network messages, and so on. The agent simply moves from one ready call to another, without stopping to dial, listen, or choose the next call.

Predictive dialing is the most powerful and the most productivity-enhancing technology to hit outbound since the invention of the list.

True predictive has complex mathematical algorithms that consider, in real time, the number of available telephone lines, the number of available operators, the probability of not reaching the intended party, the time between calls required for maximum operator efficiency, the length of an average conversation and the average length of time the operators need to enter the relevant data.

Some predictive dialing systems constantly adjust the dialing rate by monitoring changes in all these factors.

The dialer is taking a sort of gamble: knowing that these processes are in motion, and knowing that there is a certain chance that a call placed will end in failure, it must throw more calls into the network than there are agents available to handle them, if they all succeed.

Predictive dialing has been nothing short of revolutionary in the outbound call center. When operators dial calls manually, the typical talk time is close to 25 minutes per hour.

Most of the rest of that time is non-productive: looking up the next number to dial it; dialing the phone; listening to the rings; dealing with the answering machine or the busy signal, etc. Predictive dialing takes all that away from the agent's desk and buries it inside the processor.

When working with a predictive dialer, it is possible to push agent performance into the range of 45 to 50 minutes per hour. We've heard of centers going as high as 54 minutes per hour.

It's hard to imagine why you'd really want to push it that far, given how hard and expensive it is to recruit and hire agents, but it's possible to do.

How Dialers Are Changing

Like most other hardware technologies, predictive dialers are responding to changes in the nature of the call center. Agent handling, for example, is a much more complex issue than it once was. Centers that want to get the most out of their reps have to look not only at outbound talk time, but at how well inbound reps are utilized during slow periods.

To facilitate that, dialer makers have incorporated a technology to blend agents; this allows a single station to handle either incoming or outgoing calls. And although it's not used widely yet, it's growing.

The dialer is steadily losing its identity as a purely outbound object. It's got to act like, and interact with, inbound call routing systems. Because it's increasingly unlikely that a given center will be doing all of one kind of calling, or all of another.

Recent information from Datamonitor suggested that the market for outbound dialers was actually expected to increase in the next few years.

Call center managers now choose the software applications that make sense for their business (or inherit them as legacy systems), and then get cost-effective hardware to run them. Decoupling the software apps from the hardware is the most impressive development to come along in years; it was not an easy one for all dialer vendors to adjust to.

Predictive dialer vendors, like PBX and ACD vendors before them, have been forced to adapt to a changing world. People are less inclined to choose a standalone system they can't program and that can only be connected to a limited range of compatible peripherals.

How are they doing that? By focusing on their strengths. Which is the software that routes the calls, downloads the lists, tracks the results and coordinates the customer information on the backend.

If this sounds an awful lot like the new CIS software, you are right. If it sounds like computer telephony integration, you are also right.

The most successful predictive dialer companies right now-the ones making the most interesting and useful technology-are the ones that have rethought the logic of the outbound call center and recast their dialer as an indispensable component of the inbound and outbound center.

For all of them, the selling point is not the power of the dialing engine, but the value-added capabilities of the companion software.

That's why Davox bought AnswerSoft, to return to an example used in a previous chapter — because having a middleware CTI component to lay across the dialer assures the center of a solid data pathway back into the host and out to the agent's desk.

Mosaix also, is especially keen on what you can do with the data besides pump out calls at high volume.

They've recently come out with Predictive Agent Blending, which captures real-time statistics from an ACD and forecasts inbound call volumes through a CTI link.

That's not what you would have thought was the role of the dialer just a couple of years ago.

Melita and EIS are going down similar paths, with inbound/outbound blending a high priority, and coordinated data management the biggest selling point of all.

Melita, for example, has been steadily building onto its PhoneFrame platform, adding a CTI component, an Enterprise Explorer, a data navigational tool, and so on, to the extent that the dialer is almost invisible under layers of call center software.

So what should you be thinking about when buying your predictive dialer? Integration—with every other piece of hardware and software in your call center. Mostly the software.

chapter 9

Inside the New ACD

Automatic Call Distributor. Simple terms for the basic function of a call center, taking incoming calls and moving them to the right place, the agent's desk.

Over the years, things have changed. The ACD is responsible for more than just moving calls. That's probably not even the best term for it anymore; we should probably be using something closer to telephony server, though that term is already in use for the LAN server that moves call control commands from the client workstations to the attached ACD or PBX.

Now, the ACD's job is not just to route calls, but to manage the information associated with those calls as well. "ACD" is really a function that can be carried out by a wide variety of different kinds of processors.

At the very low end, you can buy a PBX that has "ACD" (read: call routing to agents) built in, or add it on through a PC application. You can route calls in the network, thanks to intelligent features built into the carrier networks.

So what is an ACD, that big piece of hardware, really for? It has matured beyond call routing. It is the brain and control point for the call center, for both inbound and outbound, for voice calls and data traffic. It's a call center's arbiter: setting priorities, alerting supervisors to patterns and crossed thresholds.

It's a remarkable machine, really, because it's managed to adapt to absorb all the technologies that have come along in the last decade: from IVR to e-mail to fax and the web, speech recognition, even video kiosks. The modern ACD is like a traffic cop; you hook on what you want, and the ACD coordinates the flow of all the traffic through the busy intersection. And tells you what is happening in real time, or what has happened historically.

I applaud the switch vendors, actually, because alone among major hardware categories for the call center, they have managed to keep their product vital and technologically exciting. Sometimes that's meant bringing technology to market that's been way out front of the call center's real need for it: skills-based routing and call blending, for two examples.

You can argue about whether call center supervisors know how to deal with the operational issues raised by these technologies, but you can't argue with their raw innovation.

These vendors, and in this mix I include Rockwell, Aspect, Lucent, Siemens, Nortel and a few others, have fought like dogs to keep up with each other in core technology, with a surprising degree of success. And while the market does segment out into big players and smaller players, into switches mainly for the large center or the smaller center, the fact is that they've all been pushing each other to stay competitive.

To wit, some of the exciting things that go on under the hood of today's ACD:

Multi-vendor interoperability. Vendors aren't selling based on the value of their hardware anymore. Instead, they are touting their software, and rightly so. Any discussion of what a switch can do will focus entirely on the capabilities of the software suite that drives that switch. Increasingly, those suites operate independently of the hardware platform.

Aspect, for example, provides ACD software that runs on Lucent, Rockwell and Nortel platforms. Teknekron has long had specialty application software that runs on other companies switches. The hardware, while not exactly an afterthought, is now "general purpose," and not very interesting. It's not where they make their money, and not what adds value to the call center.

Multiple connected servers. So if its software that counts, that software has to

35

be LAN-able and distributable. You have to be able to connect servers to what is now an ACD network running alongside or within the normal corporate network.

You're going to connect databases, fax servers, CTI servers, web boxes, IP gateways of all stripes...and as you can see, the distinction between what's a switch begins to break down rather quickly.

Which is why the choice of switch these days often comes down to which vendor is best prepared to:

- vertically integrate all the server and software combinations;

- offer the most effective tools for data management (because all those servers and multiple customer contact points make every kind of call into lots of loose data that needs to be corralled);

- and put the best consultative services on the table to get everything integrated.

MIS subsystems. The more ways a customer can contact you, the harder managing that customer relationship becomes. ACDs have always spit out reports. But now the reports are far more complex than call detail and agent performance.

Information about calls has to be correlated with data about web traffic, and back end customer histories need to be brought into the mix too. Siemens, Teknekron, Rockwell — all of the majors, in one form or another, have either partnered up or developed in house some form of advanced data management tool.

In some cases those tools are extractors, designed to let the manager at the center (or somewhere else in the company) create the kinds of reports they need.

In other cases these MIS subsystems are more for facilitating the backend work done by other, more specialized applications, like the Brightware/Rockwell e-mail routing collaboration.

In that relationship, Rockwell is integrating Brightware's software into their 3CS call control platform (3CS is an NT-based call processing system that runs on top of the switch).

Brightware's software is an internet response system; customers enter queries through e-mail or web forms, and the system can respond automatically, route the query, or a host of other options.

The collaboration is typical of the way the ACD vendors are thinking: leveraging the fact that the switches are now open platforms to hook in value-added software from promising outside developers.

Virtual call centers. A virtual center is a set of centers that perform as though they are a single unit — for the purposes of gathering management information, call data, and ideally, real-time call handling.

This unusually complex activity has boomed in recent years, as companies have tried to spread costs around in advantageous geographic areas, and as mergers in several key call centering industries (banking and outsourcing in particular) have left companies owning multiple centers, often with varying switch technology.

Virtual call center technology runs the gamut from the super-switches, like the Intecom E and their CallWise software, a platform that's hugely scalable to link multiple centers together with a single point of control, to the elegant distributed call center networks made by Teloquent, which uses the public network as its call router.

In either case, the benefits are found in the software applications; the hardware component is much less important than it used to be.

What's happening to the ACD is not a surprise; surely we've learned from other industries that hardware is always less interesting than software, that it contains fewer core benefits to the end user. That it is always driven to be more open, and once opened, the companies that can create the best value in software are the ones that prosper.

That's why we have things like skills-based routing. And switching systems — not just switches, but software-based amalgamations of hardware, software and networks — that don't just route calls. They enable transaction processing and data gathering, home and remote agents, web connections and even web-based agents.

At the end of the day, the modern ACD is not merely a piece of isolated dumb hardware, but rather an open, network-spanning piece of futuristic technology.

chapter 10

A Dynamic Trio: 3 Technologies That Are Changing Call Centers

There is a very good reason why customers demand more from call centers than they used to: call centers provide more. In recent years, companies have done a very good job training their customers to expect more in terms of service, buying options, and so forth.

This trend has been circular — as technologies are developed that make the customer experience better and more powerful, then those technologies are seen as a competitive advantage to the companies that adopted them first. Then, as customers become more familiar with the new way of doing business, they start demanding it; the technology spreads outward, and soon the bar for service is set higher. And newer technology is developed to help companies differentiate themselves.

The question managers face is identifying what technologies (or tools, or techniques) are out there in the marketplace, maturing, that will allow a company to take their service to the next level. Here are three good candidates: **speech recognition, distributed** or **virtual call centers**, and customer information systems. These are not new; rather, they are ready for use on a large scale, rather than by painstakingly patient early adopters.

This chapter is a brief introduction to what these technologies can do; in the three chapters following we'll go into a little more detail.

Speech Recognition

With almost no fanfare, this technology has made tremendous strides in the last few years. It promises to change the way customers interact with automated systems, broadening the range of telephony interactions.

In the short and medium term, the interaction of choice for a customer wanting information is still going to be the telephone. While they are migrating to the Internet in huge numbers, call centers will still be deluged with phone requests for information, service, problem solving and order taking. IVR is still the main way a caller self-routes to their information destination. Put an intelligent speech engine on top of that and you a) encourage the customer to stay in the automated system instead of jumping out to an agent and b) open the system to more users.

This technology generates a lot of excitement in the public because of its association with things like voice typing, or dialing a cell phone by voice. But clearly, the specialty applications call centers need-those that need to be speaker independent-are more powerful in the long term, with the potential to save agents time on data collection.

Consider an application created by Nuance Communications for Schwab's automated brokerage system. When I first saw this demoed in 1996, I thought it was pretty good: it understood me more than half the time, and seemed flexible. Now it's even better. And according to Nuance, it now handles half of Schwab's daily telephone stock quote volume, with 97% accuracy. With the migration of personal financial services to the Internet (and with price and service the determining factor in a competitive industry), giving a customer the ability to say "I'd like a quote on IBM" instead of typing out some ridiculous code is a key differentiator.

There are a lot of companies working on applications for this. As processing power improves and the cost of delivering a working application drops, it is likely that speech rec takes over as a successor to IVR as the "non-agent" telephony transaction.

Virtual call centers

Extending your call center across physical boundaries is, if not simple, at least a lot more possible than it was a few years ago. Many vendors of standalone ACDs saw the handwriting on the wall in the early '90s, adding features that let users link multiple centers and run them as if they were a single entity.

At first users could only link centers that were using the same kind of switch; now they can control switches from different vendors, run third-party software on top of those platforms, and coordinate the planning for an entire network of call centers.

Part of this development was user driven-when two companies merged, and both had call centers, it was important that those the cost of running those centers separately (with all their separate applications and processes, as well as call routing) not drag down the performance of the whole. Mergers and consolidations in the financial services sector, banks, and among outsourcing companies gave this trend momentum.

Today, with standalone switches fairly open but still expensive, smaller call centers are finding an alternative in "switchless" virtual routing schemes. ("Switchless" is a misnomer, albeit a descriptive one.) Teloquent's latest twist on their Distributed Call Center product, for example, uses either Centrex or PBXs with client/server networks to push calls around from site to site. The tendency to "devolve" the network from hard-wiring two mega-sites together (circa 1988) to this distributed and more flexible arrangement promises to go a step further.

Carriers have been opening their networks for several years. As the price of telecom minutes drops lower and lower, the only way for the carriers to make money on volume long distance is through value-adds-and certainly the most value you can add to a call center call is to make sure that it never has to terminate in a call center at all. Virtual call centers are coming to include everything from network queuing to carrier-based order processing and data collection.

Customer Information Systems

This is a basket of technologies from a variety of disciplines: help desk case tracking, problem resolution, telemarketing and list management, sales force automation. When all these things are combined, along with a healthy dose of the back-office essentials like data warehousing and customer analysis tools, there arises a set of software tools geared around the customer. Many people say that this is what the call center has always focused on (as have its software applications). Not so: the call center has always focused on the **call**.

CIS tools aim to eliminate the call as the metric, and replace it with customer analysis as the criteria for how you handle calls. We know that all calls are not created equal. CIS tools show how they differ, and help corporations balance what they know about the status of each call with what they know about the customer who is calling. Processes are made more consistent from one customer interaction to another. And the call center is rightly viewed as a component of a company-wide customer management strategy, rather than by itself in splendid isolation.

CIS helps change the corporate view of the call center as a place where money is spent and the only return the company sees for that money is a room full of phones ringing. It helps quantify the hidden benefits of the call center, which is that you have satisfied customers (and hopefully more of them).

chapter 11

Speech Recognition: Ready For Prime Time

Speech recognition is one of those stealth technologies — the kind that keep academics and serious R&D departments busy for years showing incremental improvement, and then all at once the development reaches critical mass and it's everywhere, in all sorts of applications.

We've had speech rec for quite some time. Every time you activate a machine using vocal tones, you are using speech rec — dialing from a hands-free cell phone, for example, or using one of the new PC-based dictation systems (Dragon Systems and IBM are the leaders in that explosive field).

The reason it is so explosive is twofold — first, the speed and power of the typical PC grew along the expected curve until it was strong enough to process speech in real-time; second, the developed algorithms were steadily improved to allow computers to discern the appropriate patterns that underlie speech, without regard to accent, speed of speech or other eccentricity.

While the PC dictation systems have focused a lot of attention in the computer media (and indeed, the popular media) on the possibility of interacting with a computer, it's clear that one of the real winners will be the call center industry.

Speech rec is starting to gain a toehold in call centers as an autoselector — a tool that the customer uses to interact with an automated system to either route himself to the proper person (an auto attendant or ACD front end) or extract the information he needs from a host database, =E0 la IVR.

It is an extender technology. By itself, speech rec doesn't add new functionality to the call center. Instead, it adds new callers: those with rotary phones, those who are mobile, those who are so pressed for time that they can't be bothered to do anything but speak. It then processes those callers using the same traditional tools that call centers have used for years. The same benefits flow from speech recognition as from IVR: fewer calls that have to go to an agent, shorter calls, and more self-service.

Moving Beyond Early Adopters

The most active users of this technology to date have been the financial services firms. These are, by no coincidence, the early adopters for a lot of call center technology, especially the Internet. That's because their businesses are focused on providing a lot of very repetitive, data-centric transactions in a very compressed time frame.

And the kinds of input that a speech rec system would have to process are very well defined-sequences of digits for things like account numbers, phone numbers, social security IDs or passwords. Or, some apps use discrete letters for getting stock quotes. There are a million ways to use it to extract information.

There are two distinct kinds of speech recognition, known as speaker-dependent and speaker-independent. The two diverge wildly in the kinds of things they are good at, and the kinds of systems needed to make them run.

Most of the attention in the world at large has been paid to speaker-dependent recognition. In this kind of system, the user trains the computer to understand the patterns in his or her own speech. By training the system over time, a user can teach it to understand a very broad vocabulary of words, and can approach 98% accuracy in transcription with certain kinds of text.

This is quite good, and it has little or nothing to do with call centers. However, this is what gets people excited-most people long for the day when they can tell their computer where to go (and have it comply).

Call center apps necessarily focus on speaker-independent recognition. Many people will call, obviously. The human brain in the form of a receptionist can recognize a huge number of variations of the same basic input-there are literally an infinite number of ways to intonate the word "hello."What you want in a call center is a system that will respond to the likely inputs-the most common words like yes, no, stop, help, operator, etc., the digits, the letters of the alphabet, and so on.

Telecom has gradually been accepting the technology in operator assistance and routing systems. (But not everywhere you think. Some automated applications that ask users for spoken input, like directory assistance, are actually just recording it and playing it for the operator, who inputs it manually-it saves time, but speech rec it isn't.)

Internationally, touch tone penetration is still very low, leaving a vast installed base of potential callers who can not access IVR. It follows that these callers are then going to be expensive to process when the come into a call center because they have to be held in queue until there's an agent ready for them-high telecom charges from the longer than average wait, coupled with the cost of agent-service (rather than self-service).

On the downside, international call centers, particularly those that serve multiple countries, can field calls in multiple languages. If you use an IVR front end to have the caller select their language then you by definition don't need speech rec. These are surmountable problems that have more do with the operation of speech rec in practice than with the underlying technology.

One More Input Channel

I think that speech recognition will be a transformative technology for call centers in the next few years.There are strong indications that speech will be part of the PC operating system, if not in Win 98 then shortly thereafter. Speech control of appliances and applications is at hand, again from a speaker-dependent point of view.

Call centers that prosper are the ones that effectively handle their paradoxical mission: reduce costs/improve service. Speech rec costs a lot to develop and perfect, but once it's done, it's done forever. The cost of maintaining it is negligible, and it has little of the headaches involved in CTI or other "fancy" call center technologies. Once you tease meaning out of the speech, it becomes input like any other, just like information entered via the web, DTMF or told to an agent.

Those call centers that add value to the customer interaction-that add traffic without adding to the cost load of the center-are of the most value to the company that runs them. The more people you can encourage to call the better off you are. Capture data about them. Learn their likes and dislikes. Try to sell them something else. And leave them with a positive experience to tell others about.

That's the essence of speech recognition in the call center-it's a simple front-end, with albeit limited application. But that's what they said about IVR ten years ago, and look where we are today.

chapter 12

"Virtual" Is Real

Virtual call centers are important because they represent a way to extend the efficiencies and benefits of the single, standalone center. Just as data networking and client/server systems wrought changes in the dynamics of how corporations perform their daily business (changes that were largely unforeseen), linking up call centers into virtual sales or customer service networks is already changing the way these operations are managed. And it is starting to save companies money.

Linking physically separate centers together is not new. The raw technology has been there for more than a decade — at least that long ago it was not unusual for a company to hardwire two centers together with T-1 lines. This was done for a number of reasons, including overflow, emergency backup, and often time-of-day coverage for companies with spread-out customer bases.

These links were expensive, and they had limitations. The connections had to be dedicated. The call had to terminate at one center and then be physically switched to the second center — costing the center twice in telecom charges, and chewing up a port on the switch for the outbound leg. And usually, the two centers had to be running the same kind of switch.

In fact, earlier in the decade the term "virtual call center" referred more to the idea of agents who worked at home and were connected to the center in real time than to multiple linked centers. There was talk of taking a "shelf" of the

ACD and putting it into a smaller, satellite center that could be located away from the main center. The idea being that you could then hire ten part timers who would work in a strip mall or some other cheap office space (saving the cost of building a new center) and run an off-shoot of your 50 or 100 seat center at a much lower cost. You could use it as an employee incentive by locating it in a suburb that was closer to their homes, etc.

This did not become a terribly popular way to run call centers. First, though it was do-able technically, it is more of a management headache than a benefit. Home agents and small sub-centers are an administrative nightmare that most call centers can do without. Second, it's not very cost effective. The business imperatives that argue for the placement of a small second center often make the case for a larger, full-fledged second center, or an expansion of the first one.

As technology improved, when the industry spoke of "virtual call centers," more and more it meant the linking of several centers together, and what they were really talking about was not "virtual," it was "distributed." Take an example. A bank has five large centers; one may handle credit card inquiries, another may be primarily responsible for handling calls in response to mailed promotions, and the others handle a variety of tasks (including outbound telemarketing).

In each case, the center has a set of defined responsibilities. The calls that it gets are probably routed through a combination of DNIS (calling different phone numbers will send the call to different centers) and caller input (press one for activate a new credit card, press seven to report a lost of stolen card, etc.). The centers could be in separate cities, countries, or right in the same office park.

What's important about this scenario is that it allows the call centers to operate as a single *unit while preserving their operational distinctiveness*. Each handles its own brand of calls, but when volume peaks, calls can be diverted from one center to another. The system as a whole can be queried for details of agent availability while the call sits, parked in queue at the network carrier level. To facilitate this, each call center will have agents trained in several of the bank's specialties, perhaps whole groups that are cross-trained.

Call flow tables will be set up in advance, just as they are in a single call center environment, so that the routing authority (be it the network or the main center) knows where to send calls according to what criteria. Managers can change the criteria in a matter of minutes, depending on the kind of system they are using. One manager told me his people can change the dynamic routing tables in about 15 minutes.

Take this example one step further. If you want to route calls based on finding the agent the most appropriate skill set to handle the call, you are going to have a lot more options when you can look across multiple centers. In the bank case, you might have 1,000 agents spread across the country. If you can identify which 50 of them speak Spanish, which 200 have the best cross-selling skills, how they rate on a sliding scale of familiarity with the company's products — all this is going to give that bank an edge in assessing the call. And even though it takes a few extra seconds to process the call on the front end, putting it into the hands of someone who can handle it better will either make the call shorter (saving money) or result in more cross- and up-sales (making money).

Skills-based routing is controversial because of the way it impacts call forecasting and workforce scheduling systems (negatively), but it is clearly a tool that will be used in an increasing number of centers in coming years. Like multi-site linking, it's a technology that's switch driven, and it's likely that the switch vendors will coordinate the ACD functions to make skills-routing and site-linking much more tightly integrated.

Management Advantages

Virtual centers not only move calls more effectively — they allow managers to staff more efficiently overall. Management information systems that span centers give them a "big picture" outlook. The routing described in the bank example lets a call center network derive economies of scale in its workforce. Some studies have shown you need 2%-4% less staff to handle the same load when you virtualize operations over several centers. At one center in the telecom industry, staff was decreased by 4% while service level was up 6%. They counted $5 million in savings from better forecasting, staffing and load balancing.

This trend will only accelerate. This is an increasingly global call center industry. It's not unusual for service bureaus to have centers on four continents, sometimes five. The cost of maintaining that kind of system is horrendous. There will be tremendous pressure to improve the productivity of every seat, and to justify the performance of every center. The technology that allows for linking centers makes it possible to squeeze more out of existing resources without a whole lot of new infrastructure investment.

I think it's going to be one of the key technologies of the next five years. And I think it's going to create a mini-industry of high-level specialists who are responsible for combining the telecom, networking and staffing resources of corporate call centers — like in-house systems integrators. And these people will be very important corporate figures in the first decade of the century.

chapter 13

Counting Customers Instead Of Calls

First, the call center was a way to answer calls. It was a way to respond to conditions that occurred not only outside the center, but outside the company — what do customers want? Building a center was a form of customer triage; the corporation needed to prove the point to the customer that if they had a problem, or wanted to place an order, the company would be responsive. As we all know, this turned out to be good for customers. So good, in fact, that it dramatically raised customer expectations, leading to a still-unfinished cycle of escalating customer demands.

Was it good for the company, too? Most definitely. Satisfied customers tell their friends, and those people turn into customers, too. Problems solved turn into opportunities. The business benefits of call centers are, by this time, old news.

What most call center professionals don't yet realize, though, is that the IT revolution that went on *outside* the call center has changed the nature of the external corporation. It is more distributed, more flexible, more networked, more data-oriented. More focused than ever before on the customer information that lies hidden in countless legacy databases.

The call center is a central information point, a library that the rest of the company can use to gather data about the relationships between companies, products and customers.

This is where "customer information systems" enter the picture. This newly-defined category of software is an amalgamation of several different software disciplines, and is supplied by vendors coming from various categories:

- problem resolution engines

- legacy and client/server databases

- business process analysis

- and networking and systems integration.

What is surprising about the category is that it has little to do with telephony or telecom.

CIS is an attempt to take all the background information that a company gathers on the people it does business with and make it meaningful. It does that through comparison, analysis, linkages. Other industries have been interested for some years in data warehousing and data mining. Call centers are good at warehousing data — the amount of information that stacks up inside a typical ACD is just staggering. What they are not so good at is sharing information in a meaningful form.

We're not talking about call detail here. Call detail is meaningful only if you are managing a call center. If you need to know how many people to bring in next Thursday, then it helps to know how many calls to expect. If you want to know who to promote, who to fire, how many headsets to buy, then the data on your supervisor station or your reports will fit the bill quite nicely.

But what if you need to know how to prioritize your calls — what kinds of callers are most valuable? What kind of revenue do they bring in? How much of your company's resources does that customer consume to generate that much revenue? If you are inside a call center and you think you know the answer to those questions, I say think again. Because chances are you haven't measured in the cost of processing the e-mail that customer sent your way. (And that you never answered because it went to the webmaster.)

Did you factor in the web site? Do have any clue which of your customers has visited it, and what they did when they were there? Do you have any way of measuring whether their visits to your website led to more calls immediately after? Or prevented calls, because they found what they were looking for? *Do you have any idea at all what your real relationship with your customers is like?*

A while back I had a conversation with some of the folks at Chordiant, who showed me some of the interesting attributes of their software, designed for precisely this application: wedding the front end of the transaction to the storehouse of information lurking behind the scenes. It's not unfair to say that Chordiant has the essential concept well in hand. Most of what we consider "external" to the call center is actually very far forward in the front of people's minds when you cross that invisible line into the rest of the corporation. And as soon as a call center is asked to cost justify itself in terms of revenue and value brought to the customer interaction, those backoffice data take on a completely new meaning.

In stepping through a demonstration of Chordiant's system, I was struck by the number of times we could frame an opportunity for the front-end rep to add value to the transaction through augmented, tailored data. We in the call center industry use the terms "cross-sell" and "up-sell" all the time, as if these practices are actually in common use. Studies show that they are not. Not because the folks who run the center don't want to, I imagine; rather, I think it has more to do with the fact that the data that makes it possible sits behind the wall in dozens of inaccessible host systems.

Now, however, we are seeing the breaking apart of old notions of who owns data, where it belongs, how you can make the most of it. In any call center interaction, indeed, any time a company crosses path with a customer, it is possible to know much more about that interaction. We typically frame that "ultimate" interaction in call center terms: how quickly did we answer the call, what was the agent performance, did that person go away happy.

But increasingly, as the web sucks us all into its e-commerce vortex, the analysis of the interaction will be framed from the other side: what did that interaction cost? Where did it lead? How many channels did that person use? Which ones were most effective at communicating the company's message, and which ones did the best job of capturing information? Those questions are now posed by people outside the call center. If the call center management community wants to be more than telecom firefighters, they will have to learn how to frame, and answer, these questions.

In some ways, it's amazing that it took so long for something as simple and obvious as "customer information systems" to develop. There were glimmers of it in the list and database management tools offered by the predictive dialing companies, and in telemarketing software, with its dynamic scripting based on customer criteria.

That's one reason we see this field so crowded with companies from all over the vendor map; every call center discipline has a role in defining where the data comes from and who does what with it. Help desk software adds its case management and customer tracking component; voice processing brings graphical call flow design tools to the table, now including web tools.

This may be the best thing the web ever brought to us — an understanding that when you look only at the telephony side of the customer equation, you get a one-sided view. With the web comes confusion, variety, and a need to see the big picture. Software that manages the company's whole relationship with the customer brings clarity to that picture. And it helps the call center do what it does best-connect with the customer.

Part Three

The Internet Cometh

The AGENT represents the company's
INTERESTS in the transaction.
Pulls or pushes the customer
towards a product,
away from a problem.
Cuts through the clutter
which inevitably results
when a company throws all its
information in the customer's
lap and says "SELF-SERVICE!"
or "SHOP ONLINE!".

Chapter 14

E-Commerce & The Internet: The Customers Are Ready

The Internet is exploding — not just in the numbers of people sharing the experience, but in terms of actual commerce transacted electronically. Though it's hard to get an exact fix on the size of the market, statistics are beginning to bear out what many of us sense: that once people get online, they gradually use the medium to explore products and companies. Ultimately, they buy.

A study by Computer Intelligence found that 30% of home Internet users were engaged in some form of e-commerce — stock trading, home banking, book buying. There is a level of trust developing on the part of the consumer. In part, it's due to the fact that brand name companies are out there, making new products available to that audience. Despite the many publicized concerns over privacy and security, consumers are increasingly willing to treat the medium as they would a phone — when you deal with FedEx, or Schwab, you are dealing with a known quantity.

And when that consumer-friendly company makes it easier to do business with them by phone, when they add value by offering services you can't get any other way, then transacting off the Web becomes a no-brainer. If I can track a package at 7am while I'm checking my e-mail, then I will. Once the consumer (or his kid) buys something by credit card over the Web — and nothing bad happens — the scare is over. They'll do it again.

At least, that's what another recent report seems to indicate. AT&T and Mercer Management Consulting released research that showed that 62% of online shoppers plan on doing more purchasing that way. Only 4% said they'd do less. Also, more than a quarter of traditional shoppers studied said they intended to buy online. And even more encouraging, 59% of consumers who purchased online said they were highly satisfied with the service they received.

So where does the call center fit in? Everywhere. The call center industry has gradually gone from being sales oriented to service oriented and now to a more balanced, integrated way of operating. Technology has made that possible. Because you can run multiple centers as one, blend calls at every seat, collate and report on calls with tremendous power. And because call centers can now manage the flow of calls coming in from multiple delivery systems: IVR, fax, e-mail, Web, even kiosks.

Web-based e-commerce makes sense for companies. It is the cheapest form of sales ever devised. It collects data, collates it, offers the widest range of product (or the most closely targeted). This is what I don't think will happen: a person, wanting to buy something, clicks on a web page to initiate a web call to talk to a call center rep. Though there are many smart people talking about web telephony and its potentials, I think it'll be a long time before that mode makes more sense for people. I think the call center will expand its traditional role in service and support, that people will call on the phone while they look at a web site, or after they've looked at an on-line catalog. I think e-commerce will be as automated as IVR is today — and as common. People won't listen to an agent or a voice unit read off a list of their account balances or stock quotes. They'll download a personal listing, then call a person if there's a problem.

I think it breaks down this way. When someone's motivated to buy something, they don't need a rep. If they'll buy from a direct response TV commercial then they'll buy from a Web site. (And Home Shopping proves that case.) But when they need help, need service, need to find out why a company does something a certain way ("why is my statement unclear about this...") then they need a person.

What vendors sometimes forget is why call centers succeeded. It's not the technology. It's the interaction between company and customer.

chapter 15

Is Internet Telephony Right for Call Centers?

Internet telephony in the call center is one of the great unsolved questions of the day. Many in the vendor community would have us believe that the Internet is now (or soon will be) a significant delivery method for customer telephone calls. Technological development on several fronts (mainly the souping up of the consumer PC and the Internet-enabling of ACDs) will transform the way companies and the call centers do business. As customers have learned to use automated systems like IVR and the Web to help themselves, so adding an Internet/phone-call component is a natural step in the evolution of consumer culture. Or so the argument goes.

I think this is a flawed notion, and that Internet telephony is not ready as a mass tool for ordering, customer service or basic customer contact. Despite some of the wonderful new tools for delivering web interactions to the agent's desktop, I think Internet telephony will remain an interesting sideline to what call centers do — process telephone calls.

There are three main reasons for this: quality, ease of use, and consumer culture. They are three sides of the same issue.

Quality. People call companies from either work or home. The average PC is always good but not great. Fire up a fast PC with some Internet-fed streaming audio today and it's about as good as listening to a crackly AM transistor radio. Try making a call over the Internet and even the most ardent proponent will admit that the sound quality is not reliably robust. Too many factors come into play: the equipment on both sides, the traffic on the net, the mediating software, etc.

Ease of use. There is simply no question that it is more difficult to conduct a conversation by computer than by telephone. Issues of proper equipment and software setup. Issues of one-call operation vs. two-call, of the number of phone lines a person has available, of closing and then re-opening an Internet connection. These are all part of the picture today. Vendors are grappling with them; some are succeeding, brilliantly. I have no doubt that three, five or ten years down the line a perfect, customer-friendly solution will be found. Unfortunately, a perfect, customer-friendly solution is already sitting on the customer's desk. It's called a telephone.

The Screen Phone?

Here is a counter-intuitive heresy to think about. Perhaps what's needed in this market, for this application, is less a drive to put more telephone into people's computers, and more to put the computer into the phone. Before I get strung up on the altar of computer telephony, remember that CTI works because it computer-enhanced a doggedly uninnovative product — the phone switch. That brought more features into the call center, including the ability to bring more information to bear on the customer interaction, as it is happening and afterward.

Internet telephony in call centers is backwards — at this moment it is technology for its own sake, without the rationale for how it makes the customer interaction better, more productive, cheaper or more satisfying. Instead of dumping more telephony into a multi-purpose PC, what customers need is more functionality in their phone. One justification for Internet telephony in call centers is that is allows rep and callers to share data on web pages or whiteboard apps. A screen phone with some data transmission capabilities will do that very well. Three observations: 1. When you buy a cell phone today, aren't you likely to buy a digital phone with built-in Caller ID, paging and e-mail? 2. Wasn't VoiceView a great idea, until it got shunted over to modems instead of helping drive adoption of consumer screen phones? 3. Is it possible that the call center industry is ahead of the curve on getting rid of the telephone set because so far, the only place where "soft" PC phones have been adopted is in call centers?

Consumer culture. No matter how popular the Internet becomes as a vehicle for e-mail communication, software distribution, promotion and marketing, I think it is extremely unlikely that regular consumers will see it as a tool for real-time conversation within five years. Not while there's a telephone sitting on their desks. Never busy, never slow. Voice quality that's always consistent. At virtually no charge.

To those who would argue that increased capacity will relieve congestion, I would point to the nation's highway system as a counterexample. More capacity leads to more traffic, and higher bandwidth applications (double- and triple-tractor trailers, to continue the metaphor).

Enough Naysaying

Internet telephony, though not a perfect call center application yet, will explode on intranets to facilitate intra-company communications. It will create and enable specialized corporate apps, like group collaborations, the likes of which we can't imagine now, just as the web did in the last three years. Some of the stuff available now is amazing, and further research and development into call center Internet Telephony could drive new kinds of technologies that change the whole dynamic.

And of course, my assessment could be wrong, they future could unfold differently, with a 1000 Megahertz PC on my desk in 2003 that can process real time audio flawlessly over a next-generation Internet backbone. Anything can happen. If someone had told you in 1985 that in 2000 your phone would be wireless but your TV would be wired, would you have believed them?

chapter 16
Web-Call Centers

Sometimes good, paradigm-busting ideas for running call centers come from outside the industry. Perhaps that's so surpassing, given that call centers have historically been very closely entwined with the vertical industries they support.

For the last two years, the industry has chewed over the many methods of connecting the Internet to the call center. Before the Web became the pre-eminent entry point, two separate vendor sectors explored the issue, from different points of view. The companies that made help desk software in the early and middle part of this decade began to incorporate e-mail into the call tracking and case management/escalation components of their system. This was due to customer demand — technical support centers were beginning to get e-mails from customers asking about open cases, or seeking general information about problems. In an effort to boost the amount of self-support customers could provide, it made sense to begin to track the e-mails coming in that could be tied to recognized cases.

At almost the same time, IVR vendors began adding Web development features to their app gen toolboxes. Although it looks prescient now, this was probably done partly as a hedge against the coming growth of the web as a self-service tool. It was known in 1994 that the Web was an excellent document distribution tool. It wasn't known that it would become a metaphor for the information age, that was just luck. But adding Web tools to IVR toolkits made a lot of sense in those days, because the flow control of a document retrieval Web contact is very similar to that for an IVR interaction, as is a lot of the backend retrieval. Only the front end changes. And in the case of the Web, circa 1994-1995, the front end was a pretty uncomplicated proposition.

Ever since then, the industry-wide conversation has focused on how to leverage the unbelievable growth of the Web and the Internet as consumer media with their potential in the call center. Most of that discussion has centered around what the available technology enables: Web-based, agented interactions, where a caller surfs the web, needs to speak to an agent, and clicks through to request it. Then, using one of a number of flow methods (one-call, two-call, voice-over-IP, etc.), that "surfer" is transformed into a "caller." The call is parsed according to traditional call center techniques with queues, hold times and agent performance all tracked according to a close approximation of what would have happened had it come over traditional phone lines.

There have been problems with implementing these tools: The technology is not mature. The Internet is crowded and bandwidth can be spotty. Perhaps most important, the consumer protocols for conducting personal business in this way are not fully developed. (They always seem to lag some years behind available technology anyway.) All these problems will be worked out in time, but that time is probably several years away. Until then, how can a company wed the real time power of the telephony call center with the web, and do it cheaply, in a technologically robust way?

As I said, sometimes good ideas come from outside the industry.

Recently I saw a demonstration of a clever twist on Web-based service. It was from a company called SiteBridge, a New York-based software development firm. Their idea is this: the Web is (for now) a text- and document-based medium. Rather than try to force feed so much information down a pipe that narrows considerably the closer it gets to the customer desktop, it makes more sense to improve the kinds of interactions you can conduct using the medium where it is strong.

Their core product looks, to eyes outside the call center, like an Internet chat system. Two users can type messages to each other, in real time. But the system, when embedded in a corporate web page, can function as an entry point into a call center where agents respond to real-time customer questions via keyboard chat. The agent has the ability to push either documents or forms back to the customer, who is viewing the corporate web page and the ongoing chat conversation in separate frames on screen.

Too simple, you say? Perhaps. But consider some of the applications.

In a technical support environment, a rep can literally carry on several assistance projects at the same time. Several weeks ago I spent more than an hour on the phone with a tech support rep while he walked me through some not-very-complicated adjustments to a laptop. Most of the time he spent was waiting for me to implement the configuration changes he read to me, and then tweaking through a dozen or more reboots. Needless to say, this was dead time for him, and I could envision him leafing through a magazine while he waited. And the toll charges mounted.

The same transaction could have been accomplished through a text-based interaction. If we were connected, he could have dumped to my screen a series of steps for me to take. While I was implementing them, he could be helping someone else, and been alerted to me by a beep or change in state of the window with my chat. Less down time, more customers assisted, and toll-free service, without telecom charges.

The system that I saw intends to give the agent automated triggers to save him typing time, combined with the access to a library of documents geared to his task that he can push to the customer.

I like the idea of this product a lot, mainly because it tackles a persistent call center objective — more calls, lower cost — with a technology that is mature, robust, workable, easily integrated, and above all, cheap. This is a bridge technology that an existing call center can implement quickly, and can cost justify based on rep efficiency and toll reductions. And it offers a way to enhance the perceived reach of the web site (all the way into the call center) without forcing you to use technology that's not ready.

Because this good idea came from outside the industry, there are some call center specific needs that aren't completely addressed in the first iteration, things like distributing the call to an agent based on specific criteria, and tighter back-end data and ACD integration. I'd like to see this combined with e-mail and wrapped into a customer information system, for example. SiteBridge's products and those of its competitors will probably evolve based on the specific customers that use it early on, and it's likely that the smallish companies developing these kinds of systems will be heavily influenced by the needs and desires of the established call center and telecom companies they end up partnering with as they enter this market.

But it goes to show that sometimes good ideas rely on simple, existing technology implemented in a new way. And sometimes it takes someone from outside to give you insight into how to run your shop.

chapter 17

The Internet in the Call Center: A New Model

Few would have predicted that the internet would in so short a time become so important to business. Has there been any other ten year span in which the basic business tools at an average person's disposal have changed so dramatically?

In 1986 I was hired by a major media company (since merged twice into a multinational media giant). They sat me down in front of a typewriter. They had no fax machine, no multi-line phones. No PCs. The only state of the art piece of equipment was the copy machine, and like the networks of today, that had its own priesthood class of people who were allowed to use it — everyone else had to go through them.

The company's model for customer service was similarly quaint. Suffice it to say that there wasn't one to speak of, and that wasn't unusual in 1986.

What the internet has done, I think, is in one sweeping wave change all the rules for conducting business. There are now alternatives for how to work, where to work, how to communicate, how to inform oneself and one's customers. The customer too has choices (maybe too many). Surely there are many people and companies operating on the old model, but they are now being replaced because they are not competitive.

The very idea of a call center is revolutionary in this context — call centers are automated service delivery points, chock full of data about customers and products, dedicated to one core function: provide the customer with whatever he wants.

It helps to do it cheaply and fast, of course, and that's where all the attention has gone in call center development in the past few years. And that's where the internet has come into play.

An Agent In Every Home

With an internet connection on every desktop at home and office, each customer has the moral equivalent of the agent's terminal sitting in front of them. Why use the agent as an intermediary for things like account balances, or brochures, when a customer can get that stuff himself?

The reason you still want (and need) an agent present is inherent in the very word agent. The agent represents the company's interests in the transaction. Pulls or pushes the customer towards a product, away from a problem. Cuts through the clutter which inevitably results when a company throws all its information in the customer's lap and says "self-service!" or "shop online!".

The agent is also the representative of the customer, when there is a problem. As the point of entry, the customer needs someone who can guide him throw confusing options, through procedures, and who can assure the customer that their needs will be taken care of.

Models For Integration

So far so good, but how do we integrate the internet into a call center that's been designed from the ground up as a telephony contact center? Several models have been proposed.

The first steps toward integrating the call center with the internet came in the help desk. E-mail was a natural way for technical support customers to register their problems and track them as they got resolved or escalated.

Also, the internet provided an early tool for distributing technical documents to a wide community of problem solvers, sometimes including the customers themselves. This was about five years ago now, and the web wasn't yet a factor in planning for the call center.

The model still survives, though, and forms the basis for most real life combinations of the call center and the internet. E-mail, used by the customer for communication, still forms the bulk of non-telephony interactions in the center.

One problem that this created, which still exists, is coordinating among the multiple streams of input that you've now put into callers' hands.

What if an unhappy customer sends an e-mail to open a case, and after an hour or so without a response, calls an agent and starts a new case? The current generation of customer information systems can sort that out, but early on this upset service statistics and made life difficult for the help desk manager.

As technology became more sophisticated, and particularly as IP telephony was developed as an alternate mode of carrying voice traffic, the predictions for what a future call center would look like began to border on the bizarre.

One model still in current vogue involved customers who click on web sites to request either a call back or to initiate an internet-based phone call with a call center rep. The logic behind this is sound from one point of view: the customer is already armed with information about the transaction.

If you can get the backend straight, getting the agent in the right place with info about the customer, and if the telephony works perfectly and the call is placed within a reasonable time frame, then you have a good interaction.

The jury is still out on whether you will have a cheaper interaction, or a better one, because we don't know enough about how this model works in practice in high volume.

The products that make this kind of interaction happen are fantastic examples of cutting edge technology that really push the boundaries of what's possible. The trouble is that the market hasn't found a way to integrate them into systems that we know work - the daily grind of call center call handling.

Real-Time Text Connection

Another, interesting model for connecting the center and the internet was described in the last chapter — more of a real-time text-based interaction. In this scenario, a caller connects to a web site and instead of asking for a voice call, asks for an agent's assistance that comes in the form of a chat window.

If you remove the hardware and bandwidth necessary for carrying on a voice call from the equation, you allow for an interaction that is live (or appears live to the caller).

The rep sitting back at the center can handle multiple callers at once because of the delay inherent in chat mode, and can use scripts to speed his responses.

Also, the rep can guide the caller to a particular web screen, share information and participate fully in bringing that transaction to a successful close.

One of the major benefits of this latest model is that it moves the internet/call center connection from the service side onto the sales side, and does it at a technology level that doesn't rule out the smaller company. It allows you to bring in reps that can guide a web surfer to a sale (or whatever the next level of the sales process is) with the same cost savings associated with the more intricate "call me button" model of web interaction.

I like this model a lot. Companies that focus on integrating chat and e-mail into the call center, like SiteBridge, Mustang and Brightware, are the next wave of strong application developers for call centers, akin to what the middleware vendors did in the last five years.

Some see this as a transitional technology. That is true, but as we look back at the history of the call center, it's hard to see when the dominant model for providing some form of automated sales or service wasn't in some form of transition.

I think that at some point in the medium-term future the call center will be connected to customers in all sorts of odd ways, including through some pipeline that either the internet or its successor.

Although at the moment there appear to be many ways to effect that connection, the choice of a model is going to depend on the resources available, the comfort level with transitional technology, and the relationship between a company and its customers.

Part Four
Call Center Ops

As we add more ways
for the CUSTOMER to come
in the front door,
the CALL CENTER remains
the key customer TOUCHPOINT,
no longer isolated
from the company
by technology—or
business strategy.

chapter 18

Fax: The Forgotten Process

With all the attention paid to the new, high-tech pathways into the call center (video, Internet, etc.), it is easy to forget that the most popular way of making contact with a company, aside from a plain old-fashioned phone call, is by fax.

The press of paper-based transaction information is still a huge part of the process inside call centers. In a sales environment, faxes are a key ingredient in the confirmation of transactions. Many customers insist on getting a fax as part of a complex transaction like booking a reservation, or applying for a loan. And on the inbound side, there are countless ways that fax becomes part of the equation — order-taking, warrantee registration and simple requests for service or assistance, to name just a few.

"What we've learned is that most call centers view fax as a necessary evil — they don't like to deal with it as a message medium, but they have no choice," says Jerry Rackley of Teubner & Associates. "The constituency they serve prefers it. The result is that fax volumes are growing in most call centers and often these faxes are handled like mail once they're received."

Unfortunately, these faxes are not usually handled with the same kind of sophisticated automation that accompanies other call center-based transactions. As volumes grow, the pressure to deal with this mounting pile of paper increases. "We have customers who had 40 or more fax machines in their call centers to accommodate the load," Rackley says. "Why haven't more call centers exploited fax automation to rid themselves of their fax headaches? I think the answer is simply that they don't know that they can. Fax technology, despite how easy it is to justify and implement, has maintained a relatively low profile in the marketplace."

A recent survey of call centers conducted at the Incoming Call Center Management tradeshow by the Call Center Network Group revealed that 72% of call centers send or receive faxes as part of the caller contact process, but only 28% have automated that process.

In the high-volume environment of the call center, the documents being faxed or received are often generated by an application, not a user. Teubner sells the Faxgate system, typically configured with 16-24 lines. Rackley says that they often are attached to multiple platforms, like an MVS host, an AS/400 and a LAN. Volumes of documents being faxed often are several thousand per day. These systems sell for $24,000 and up.

"We frequently sell two identically configured systems, one for primary and the other for backup-an indicator of how mission-critical the documents being distributed through these servers can be," he says.

"Our call center customers are using fax technology to receive credit applications, purchase orders, claims forms and to distribute sales quotes, bills of lading and other documents. In all cases, they are saving time and money as well as gaining some sort of competitive advantage."

Given that fax is going to be part of the call center picture for the foreseeable future, those organizations that use it should apply the lessons learned by automating other processes. Key is investing in a high-volume, LAN-based fax processing application that will track and route the fax information without adding to headcount.

In fact, from the manager's point of view, the less interaction the agent has with the fax the better off the call center as a whole. The agent's time spent printing, faxing, retrieving faxes, and following up should amount to very close to zero.

All of these things should be handled by a center-wide system tied directly to the applications running on the agent's desktop.

If, in the course of a transaction, it becomes necessary to fax something to the customer, the only thing an agent should have to do is collect the fax number (if you don't have it already) and identify the information the customer wants faxed. The application should do the rest.

Call centers that are still running on manual are missing the boat-the technology to automate fax processing is far more mature than anything involving the Internet, and is today more of an "enterprise-wide" application than most of the software that purports to do this.

This is a relatively simple technological solution to the productivity problem-too many pieces of paper, too many agents chasing down too many loose pieces of information, too little efficiency.

Call centers that are looking for a quick productivity boost should look no further than the fax double-whammy: fax-on-demand (for getting information to customers from behind a voice front-end) and fax servers for hands-free bulk information collection and distribution. These truly are mission-critical applications.

chapter 19

Standardizing Business Processes

As the call center evolves from a simple telephony contact center into a more complex organism, its functions change, and so does the way it relates to the rest of the organization. Much has been said about the new kinds of technology that link the various parts of an organization — indeed, for much of the last ten years we have watched as a long series of enabling technologies were first envisioned, then developed and finally made available as mature products for connecting the call center to its data repositories.

In some ways, questions of technology are, if not irrelevant, at least beginning to take a back seat to some very important procedural issues. Companies are beginning to focus their attention not on the call center itself, but on the call center as a component in a company-wide strategy of customer acquisition, management and retention. The business value of a call center is not that it helps keep costs down. It should not be measured in calls answered or not answered. Instead, the unit of measurement that makes the most sense these days is the transaction: whether a person reads a catalog, calls an 800 number, hits a web site, or some combination of these and others. The goal of the successful customer management strategy is to make sure that all these parts work closely together to ensure a smooth relationship and maximum information value in both directions.

You have to make sure that every time a customer touches your company, the information that's gathered is processed in the same way. No duplication of effort. Say a person needs information that's not readily available — the answer to a tech support question. No matter how the customer comes into the system, whether he posts an e-mail or leaves a voice message or actually talks to someone, that interaction needs to be tracked in exactly the same way. Now, ask yourself, what are the chances that a given company has the data management tools, and the pipeline between call center and IS, to make sure that actually happens? Isn't it more likely that there are separate databases, separate transaction–reception interfaces? Isn't that e-mail likely to be handled by someone far removed from the call center? Isn't it likely not to be tracked at all?

The call center industry has paid a lot of attention to the details of the call. There are more ways to analyze call detail reports and ACD stats than there are call centers. This is something people have learned, over the years, to do. And there are professionals outside the call center, in marketing, customer service and product development, who have made careers out of taking those call center stats and applying them to their own disciplines. Inside the call center, hundreds of automated processes have been created for internal and external procedures. One of the reasons the call center has flowered as a corporate component in recent years has been because it has embraced automated processes as a replacement for manual — and ad hoc — transaction processing and decision making.

As the physical call center becomes one of several avenues into a company, it becomes more important than ever that companies model and coordinate their processes to bring them in line. There are two key benefits to process modeling.

1. Enhanced flexibility. In large companies, there may be hundreds, if not thousands of defined processes for dealing with customers. These processes may range from the simple (collect certain pieces of data each time, like name or customer number) to the very complex (analyze the customer's history, arrive at a credit score, and based on that, offer the customer a certain upselling opportunity). Why should the process be different if a person is calling via web or telephone? Clearly, if your corporate process says that a customer is right for a certain upsell, then he's right for it. Because you have no control over how the customer comes in, coordinating the procedures to operate across the available entry channels is critical. And when you change procedures, you need to have that change reflected across all the customer interaction channels. Why should the website still be offering a customer yesterday's product at last month's price?

2. Coordinated process modeling helps you keep control of costs by adapting your processes to the customer channels that make the most sense. One expert told me that the average cost of processing a telephone call was $5.50, and that the similar processing cost of a web interaction is 25 cents. Defining the customer interaction processes helps a company determine how to handle customers in the most cost-effective way, offering them incentives to use the less expensive transactions, yet ensuring that the kind and quality of service they receive is consistent from one channel to another.

With thousands of processes to model that are constantly changing like so many strands of spaghetti, how does a company impose a sense of order and consistency? For the largest, the best answer may be to bite the bullet and have a huge accounting or consulting company come in and blaze the path. Mid-sized companies, though, can turn to the growing field of *customer interaction software*, a new area forged out of a combination of help desk systems, telemarketing/scripting software and CTI middleware.

Some of the vendors in this category are making "process management" a key element of their offerings, perhaps recognizing that unless the call center coordinates better with the rest of the organization (especially with IS management) it won't be able to offer the organization the cost advantages and flexibility it did when it was the only game in town.

chapter 20

Call Center? Or "Customer Touchpoint"?

Every time we talk about the call center, we are using a shorthand. The "call center" industry no longer represents just telephony. It is instead a point of business contact that encompasses so much more. And as the definitions stretch and recombine to include so many more technologies and what are probably best described as "customer touchpoints," planning for call centers relies more and more on understanding the role of the call center within the organization. And on understanding the customer — who he is, what he wants, what is his value to the organization.

"It's a business center, not a call center," says John Palmer of Chordiant Software. "It's a place to transact all aspects of the customer interaction. The call center has become a lot more important than a lot of companies envisioned. It's the place where customers come into the organization. And not just through the telephone," he says.

Palmer makes the point that today's customer interaction is, on the surface at least, very similar from the way it was in the past. Call centers are, after all, still 99% telephony-driven from the customer point of view. Agents account for the majority of contacts, with 10-20% sliced off by a voice response system. "But that's dramatically going to change because of the enabling technologies," he says, as companies make better use of everything from automatic teller machines to video, e-mail, fax-on-demand and, of course, the Web.

"It's really apparent to us that the call center has to be a virtual center for all customer touchpoints. A company can't have e-mail going somewhere else, can't have the Website hosted by some other department. It needs to reside in one virtual system," Palmer says.

What that means, in practical terms, is that the way the call center relates to the rest of the organization will change somewhat. The front-end is progressing nicely. Customers are becoming used to some of the more exotic contact systems, and are starting to choose them for particular kinds of interactions. E-mails for support questions, for example; Web for financial transactions, that sort of thing.

Call centers have done a pretty good job mobilizing their data. The collection of customer information has been handled technologically very well for years. CTI has enabled a two-way passage of information so that the most important data arrive at the agent's desktop as well. The ACD is as much an information appliance to the modern corporation as any other network server. In that sense, mission accomplished. All that remains is for prices to drop on the technology, and that will surely happen.

The one area that hasn't really been addressed well has been the underlying business logic of the customer interaction process. And why should it have been? Until very recently, the act of handling a customer has been largely an act of triage. Make these calls go away! And make it cost less to do that! All the key business issues were easy to deal with — indeed, easy to ignore — when the only customer touchpoint was a phone call. Count the calls, plot out the raw costs and productivity involved in handling them, and voila, you could get a quick taste of the value of the call center to the organization.

But what if you have to add into the customer equation countless thousands of Web hits that go unanalyzed? How do you create a strategy for acquiring, keeping, and managing customers without acknowledging that all these touchpoints exist, have different cost structures, and provide different results for the customer (and the company?)

Some suspect that the cost of handling a typical telephone call is somewhere around $5.50, while the cost of a Web-based interaction is close 25 cents. With that kind of disparity, you have to explore the medium, but you can't explore any medium by itself anymore.

Planning for the future requires that a company create strategies, implement policies that are, in one sense, removed from the technology. Do not think about how that customer will get to you. Instead, know that hiding somewhere in the legacy system is truckload of data about all your customers. All the hype about data warehousing and mining is real, for one very important reason: if you do not know who your customers are and what they want, you will fail to keep them, and you will spend obscene sums trying to get new ones.

Palmer cites the example of a financial services company — they may have hundreds of separate product offerings, dozens of ways to profile customers, and thousands of unique processes for employees to follow in connecting those products and customers. "We can model them in software," Palmer says, " and then automate the processes." Once they are documented, they can be adjusted, adapted, reshaped to fit new business models or new kinds of products. Changes can be executed in months, rather than years.

"You can look at the call center from the point of view of 'cost points' and 'value points'," he says. Once you capture the information from one customer channel, say the Web, you can collect a consistent data and statistical set. "The whole environment operates at the task level — you can create a database of all the workloads and events at the task level. Not just 'Did it take two minutes to handle the call?', but what actually happened during that call."

The vendors in the emerging category of customer information (or interaction) systems are paying a lot more attention to systems modeling and process documentation than anyone ever has in the modern history of call centers.

Finally, companies that come from a variety of starting points (telemarketing/scripting systems, customer tracking systems, help desks, even CTI middleware and consulting companies) are approaching the call center from a point of view that does not start with telephony. It starts with the customer, as represented by the piles of customer histories and interaction processes that sit inside every company.

I think it's a sign that the call center is a healthy and expansive part of the modern company. It remains as the centerpiece of any smart customer contact strategy, but it has to be augmented by the huge resources of the rest of the organization. As we add more ways for the customer to come in the front door, the call center remains the key customer touchpoint, no longer isolated from the company by technology — or business strategy.

chapter 21

Confronting Disaster in the Call Center

Note: This chapter is being written in Spring 1999. With that caveat out of the way, we can speak of the potential for disaster stemming from the advent of the Year 2000. If you're reading this post-2000, congratulations. Hope everything went ok.

Despite assurances from vendors and government agencies, the turn of the Year 2000 will be disruptive. How disruptive we don't yet know.

You may be optimistic, or pessimistic. Nowhere in this article will I tell you how bad I think it's going to be. I'm not sure. But I will say this — if I ran a call center, I'd be prepared. If you're a pessimist and you turn out to be wrong, you get laughed at. If you're an optimist and you're wrong, you go out of business.

Let's recap. The problem is that many computer systems are unprepared to cope with the calendar's rollover from 1999 to 2000, for a variety of reasons. There are some other, related problems ahead as well, like the possible failure of the Global Positioning System in late 1999, and the kooky fact that 2000 is a leap year, which 1900 wasn't, resulting in a lot of confused computers come February 29th.

What many in the pessimist camp are predicting is a series of system failures across society. Telecommunications; government tax collection and transfer payments; banking and financial services; the electrical grid — all these primary systems are vulnerable to the failure of date-sensitive software. And then there is the ripple effect: a system that breezes through the rollover without a problem may fail because it relies on one that wasn't compliant. It all depends on how tightly interconnected your technologies and processes are. As I write this, hundreds of thousands of workers across North America are not making cars, largely because of a strike at just a couple of plants in Michigan. Much of the system failure has to do with the fact that parts shipments and inventories are controlled very tightly to reduce costs. When one link in the system stops functioning, the whole grinds to a halt.

Let's stipulate that failures in banks, tax collection, Social Security payments, water, electricity, nuclear power, and transportation are all much more important to the world at large than whether the call center industry functions in January 2000. I simply want to point out that call centers as we know them might cease to function. The only way to avoid that is to prepare now.

Call centers rely on a dramatically fragile web of technology to operate every day. The telecommunications network, local and long distance. The electric grid. The ability of staff to get in to work every day (telecommuters notwithstanding). The ability to boot the computer and log into the network. To process orders, fulfill those orders, issue confirmations, track requests, and so on in an endless list of daily functions that have accumulated over the last ten years or so.

The call center press (myself included) has historically made a half-hearted effort to remind people every now and again that disasters happen. There's the annual coverage of the call center that gets flooded, or that has its calls zeroed out by a cable cut or network outage. These are presented as warnings, and the prescription is always the same: uninterruptible power supplies, power protection devices and disaster recovery contingency planning.

And in ten years of talking to call center managers, I've never had anyone tell me that a) they have a disaster plan, b) they drill on the plan regularly and c) the people who are counted on in the plan know what their roles are and how to execute them in a crisis.

Just the opposite, in fact. The incidence of "disaster" has always been so rare that warnings about planning fall on deaf ears. Sure, some unlucky center is put out of operation when Chicago floods or an earthquake rocks Northridge. But in such a big industry, it's always someone else's center that gets hit, and we all end up feeling better — not chastened — knowing that we dodged a bullet.

The Year 2000 problem could be different, because this time the problems could very well be systemic, affecting entire regions, market segments, in some cases entire countries. It's possible that every call center whose ACD runs on a certain brand of PBX might fail to operate correctly. It's possible that all centers within a utility's service area might lack power. Yes, you can triage calls and answer phones without computers, but will people come to work without heat or air conditioning?

If there are major product problems from Y2K fallout in other industries, it's the in-house call center that will feel the pressure from annoyed consumers. Will out-sourcers be in any position to help clients if they can't help themselves?

There are so many variables that it's almost impossible to quantify and plan for. There's a lot of talk about what's going to happen, and almost none of it informed by facts.

So what should you do? There are some precautionary steps to take. I have no advice for how to mitigate the real effects of a real Y2K failure. But I do know that no one has any right to be surprised by those effects, not with 18 months to investigate.

First, take stock. Inventory the critical local systems. Contact every vendor that has supplied hardware, software or critical services to your call center. Because of the ripple effect, this has to be a companywide effort, and has to include the systems connected to but not part of the call center: essentially anything IT. The goal of this exercise is not to fix anything, but to get a sense of how vulnerable you are internally. If your premise systems fail internally, how reliable are those who can fix them? Get your vendor on the phone and find out what exactly they mean by Year-2000 Compliant. Ask yourself: If this system fails, how am I hurt? What will I do instead, if I can't have it working for a day, a week, a month, six months?

Take stock externally. How redundant are your telecom systems? Do you route or queue calls in the network? The most important thing you can do is find out where the weak links are, so that you can either jettison them, or work around them.

Which brings me to the third thing you can do. Somewhere in every call center is a person on who's desk sits a dusty black binder. It may be labeled Business Continuity Plan, or something like this. Find this person. Examine this binder.

Chances are it was never tested. Test it now. See where it failed. Rebuild it. Show the results to the highest rung on the corporate ladder that will listen. Get resources committed to bulletproofing that plan. Because you work in a call center. If you can't handle calls, you go out of business.

The call center is so closely tied to the health of the modern company that it's not hard to make the case for testing and implementing a continuity plan. Think of it this way. You can't rely on the rest of the world's systems to function properly in the Year 2000. But in a call center, if all you have is phone service and people at their desks, you can at least make a go of it. Put a plan together to assure that, and if disaster strikes you'll look like a genius. You have 18 months. Go find out how screwed you are.

chapter 22

Losing A Lifetime Customer: How The Call Center Can Become The Single Point Of Failure

I don't like to use personal experience as a platform for making generalizations about the call center industry. But I had an experience with a center recently that might prove instructive.

In dealing with a financial institution, a major bank, I had reason to question several items on a statement. Two of the transactions listed were mysterious to me — they were debit card transactions that listed a date and a vendor address, but no other information. I called the bank's help line to find out what company was behind those charges.

The call flow progressed normally — an IVR system answered and presented me with a seven option menu, plus the choice to *0 out to an agent, which I chose. The system then asked for an account number, which I entered.

Since this was the early evening, I wasn't surprised to find myself waiting in the queue for 10 or 15 minutes before reaching an agent. Since this was a local call, that's revenue-neutral for both the bank and the customer.

I explained my request: can the bank tell me the identity of the vendors behind the two questionable debit transactions? The rep checks. The first answer: the only information that's available is what's on the statement.

After several minutes of discussion, the rep finally tells me that because we are talking about debit transactions, which are a fairly new consumer offering for the bank, there is no information in the system that she can extract beyond what I already have.

I attempt to escalate, and ask to speak to a supervisor, and after a few more minutes, one comes into the conversation. She has some information about the problem, but she is unaware of the details. From this I assume that the rep conveyed the problem to her verbally, but that the screen-based account information that the rep was looking at was not transferred to the supervisor station.

The supervisor's response to the problem was two-fold. First, she attempted to explain to me why the information I wanted was not the bank's responsibility (in her words, because of the nature of debit card transactions). If I wanted to know who the vendor was, I should go to the vendor's address listed on the statement.

I knew that no vendor I do business with was at that address, or I wouldn't have called. I was concerned about fraud, stolen card numbers, mistakes, all the things that the bank is supposed to help their customers deal with. Instead, the initial response was to foist the question off on the customer and refuse to deal with it.

Her second response was even more interesting. The only way I could get more information out of the bank, she said, was for me to disavow the charges, sign an affidavit alleging fraud, and have the bank begin an investigation. The bank is telling a customer that the only way to perform due diligence on an incomplete transaction is to swear out a legal complaint against a possibly legitimate vendor. And the cost to the bank of pursuing an investigation (just in paperwork alone) has got to be higher than the cost of having the agent find the information.

Ultimately this was resolved when I discovered, long after that call, that the vendor on the transaction had incorrectly set up their point of sale terminal — were reporting the wrong address to the bank, which passed it on to me. The questionable transactions were legitimate.

If I had followed the bank's two pieces of advice, I would have 1. gone to the wrong address to find the vendor and then 2. charged a decent local vendor with fraud. Time, expense, frustration for all three parties to these simple transactions.

What lessons do we learn from this?

Anticipate the unexpected. Customers will call for all sorts of reasons. Agents and supervisors must be given the tools to delve into customer records and transaction histories. They must be empowered to say simply: " will look into this and I (or someone else) will get back to you at this specific time."

The agent must have a customer-helping position of last resort. Too often that last resort is negative: I can do nothing for you. Instead, it should be positive: We will take these concrete steps to look into the situation, even if it can't be resolved right now.

Even if the solution is just a stall for time, it has to be better than how this supervisor handled it.

Implement clear, streamlined procedures. You would think that a customer call to inquire about a statement would be a common occurrence. However, that simple informational transaction was not an option on the IVR menu. And it was not part of the standard roster of options the agent was initially equipped to deal with. She didn't have any more information than what I had in front of me — and no knowledge of how to interpret the bank's own codes that were the key to understanding the statement.

From the supervisor side, the escalation process should have been clearer. There should have been a way to investigate a transaction without forcing a customer to assert fraud where none exists. That's a supervisor's way of frantically reaching for a solution because her management hierarchy never anticipated my question, and therefore never provided her with a solution.

As an industry, we tend to get focused on how alternate contact channels serve the customer by giving them access to their account information, etc. In all the talk about call center/web combinations, kiosk implementations and so forth, it's easy to forget that those tools are a) better at selling than servicing and b) merely extensions of the essential human interpretive interaction that most customers want when they need help.

It's baffling that a bank wouldn't have a clear business process for dealing with customer statement inquiries. It's strange that they wouldn't have a straightforward step-by-step process for leading the agent through this kind of interaction. Perhaps they are too focused on the fact that the customer service transaction has no hard dollar reward for them, or that too many of the customers who call in for help are low volume, little profit.

Given the state of their call center technology, as I assessed it during the call, I don't believe they make an assessment of the lifetime value of the customer before they figure out how to handle the call. They have limited CTI capability (limited to primary screen pop), and little backend database connectivity to the agent desktop. There doesn't seem to be very much data automation (as opposed to call control automation) at all.

We as an industry also tend to view the financial services sector as both a ripe market for call center technology, and a leader in its implementation. That stereotype is not always accurate, at least on the customer service side.

chapter 23

Monitoring: What Price Quality?

Monitoring agents is often considered the best way to ensure quality service. Some industries, like financial services and insurance, use recording and monitoring technology to provide a verifiable record of a customer transaction.

There's no doubt that the techniques of random and total call monitoring are widespread and considered part of everyday life in the call center industry.

That doesn't mean they are totally accepted by agents. Or by managers looking for ways to motivate and retain talented agents.

In an environment where turnover is high and so is the cost of hiring for a large center, an agent-friendly monitoring plan is more important than ever. It is possible to implement one without hurting employee morale.

You can come at quality from two directions:

- making sure you have the best agents possible, operating at the highest level they can personally achieve;

- and enforcing a consistent standard of quality for customer contacts, from the customer's point of view.

Conventional wisdom says that you shouldn't train a rep without listening to his or her phone technique.

Monitoring is a critical part of the process of teaching a new rep how to deal with customers, how to handle difficult situations, even how to follow a script and read a screen full of complex information.

Feedback is important. Without it, reps don't improve. Even reps that have been on the phones for some time need constant skills assessment and further training. That's just common sense.

Monitoring agents is still the best-tested way to ensure that you achieve quality from both standpoints. The knowledge that a random selection of their work can be taken to represent the whole is a powerful incentive to "behave"–that is, to perform within well-established limits, not to deviate from standard procedure.

But this causes a tension that could have two potentially devastating effects. First, over time it can cause to occur in a center a process like natural selection, tending to favor agents who adhere to the central norm. You might be selecting against agents who are particularly creative, who (to use a tired but telling phrase) can think out of the box to solve problems.

Clearly it's not just the monitoring that causes this; the monitoring has to be paired with a too-strict behavioral code for phone work. And it has to be enforced in a way too draconian to allow that kind of agent to prosper.

(You might be reading this saying to yourself, "that doesn't describe my center's situation," but are you sure?)

Those agents that find the pressure too stifling leave for other jobs, often similarly skilled work (in service industries) but not in call centers. This is not surprising; who, after all, are the hires? College students, younger people, part-timers, people who are comfortable with the odd-shift and transient nature of call center work.

Within those groups there are large segments of people for whom monitoring, especially badly implemented monitoring and feedback programs, are an invitation to take a hike.

I'm not blaming monitoring for high turnover, but I am saying that an industry that wants to reduce turnover and make hiring cheaper, have people stay longer, should think hard about the mismatch between the kinds of people it hires, the kinds it will need to hire in the future, and the way job performance and behavior is measured.

Which leads to the second problem that monitoring indirectly leads to. If the role of the call center agent is changing over time (few would disagree), and that change is accelerating (due to the impact of the internet, e-commerce, and an altered relationship between companies and their call centers), then the agent of tomorrow is going to need a broader skill set than the agent of today.

That skill set is going to have a lot less to do with how well they use a computer or how many languages they speak, than with their interpersonal and problem solving skills.

Many companies pay lip service to the futuristic notion that a call center agent will be the "ombudsman" for a customer relationship with the company. That all basic interactions will be handled automatically, and the only customer interactions that will be agented will be those that require the human touch.

Well, that human touch is really a combination of empathy and hard-headed problem-solving that lets the customer feel like he's being taken care of. But that also represents the company's core interest, which is to present successive revenue generating opportunities to the customer without seeming crass. Talk about a skill! People who can do that are the people who, I think, come up the worst in a bad monitoring environment.

Good Monitoring Or Bad

If handled with sensitivity, monitoring can be a benefit to agents because it helps them define and reach career goals, assess strengths and weaknesses, and mark their progress according to realistic standards. But it does impose a degree of stress.

For centers that use ACDs, it's more than likely that you'll have a monitoring system built right into the switch. For smaller centers, though, you may have to look outside your phone switch for add-on products that help you monitor.

Some cities and states already have rules that may affect how you monitor employees. All call centers should fully check all local regulations before embarking on a monitoring program (or purchasing equipment).

The most obvious benefit of monitoring is that, if done right, it creates an objective standard of behavior that can be measured, and if found good, repeated. It helps ensure that you deliver not only good service, but consistent service. Consistent from each agent, and consistent across agents.

From the agent's point of view, it creates a way to measure performance that can be spelled out in advance and critiqued intelligently. Results can be quantified and reps can see improvement over time. Plus, it allows management to benchmark standards and ensure that all reps are treated fairly, by the same standards.

Experts in agent management say that the key seems to be to inform reps about the process, how it works and what it is meant to accomplish.

Also, use the monitoring program for feedback, both positive and negative, so agents have an opportunity to explore with managers what happened during recorded customer interactions. And perhaps most important, establish objective, scorable standards for performance, and state the reasons behind those goals.

There are powerful reasons to use monitoring as part of a quality assurance program. Consistency of service is one of the holy grails of the modern call center, rarely if ever achieved.

But if call centers are going to use it, they have to use it in a way that's not going to alienate the best agents, and that's fair to all agents.

Part Five

Call Centers & The Wider World

CALL CENTERS are in existence because companies have recognized the **IMPORTANCE** of **CUSTOMER CONTACT**. Economies are in existence for that very reason, too.

chapter 24

This is a Global Industry

This is a harshly interrelated world. National economies, vertical industries, stock markets, all are integrated to a degree unheard of just a few years ago. From one perspective, the world is a single market, and we are all customers of, and suppliers to, that market.

Nowhere is that clearer than in call centers. Just look at outsourcing. This year the market performance of the public companies that provide call center services have taken a beating. Last year, they were Wall Street darlings, but the market had what I would describe as an impure understanding of the hows and whys of delivering call center services. I don't think the outside world understood just how commoditized call center services had become.

Think about how low the barrier to entry has become for call center services: you need a site with good telecom. That narrows it down to just about anywhere in the US or Canada. Is there any location that doesn't have adequate telecom bandwidth these days?

You need the fundamental equipment. ACD, networking, call center software. If your aim is speed, you can pick that up from a turnkey provider. You can actually get an end-to-end call center from one of the umbrella suppliers for a 10-15% premium over buying the components separately. For the price conscious, these infrastructure items can be had on the secondary market just as easily.

And more important, most of America's top companies already have these assets sprinkled throughout their enterprise somewhere. If a major car rental company, hotel, bank, brokerage or HMO decides to open a call center, it's increasingly likely they can draw on existing internal resources (including managerial leadership).

It turns out that the hardest part is assembling the phone staff. Yes, the labor pool is still the wild card in the equation. It's very location dependent, and the skills you need are very specific to the kind of business a company practices.

Since anyone who wants to can get into the business, many are. It's no longer unusual for a company with in-house call centers to consider outsourcing their excess capacity. The industry has assumed a growth rate of 15-20% per year for the last few years (a statistic I won't vouch for, but merely repeat). It's clear that that cannot continue, and the anecdotal evidence shows that there is already a slowdown in the rate of increase in new call centers in the US. Other areas are picking up the slack. And in large measure, those areas are across borders.

The rest of the world is catching up to the US in its call center sophistication. Last spring I got a call from a gentleman who had no call center experience, but had a pile of money. He was interested in starting an outsourcing business. He planned on simply assembling the components, opening the doors, and grabbing a small piece of an outlandishly large market. I expect he will do just that, and that he will have some marginal success.

Now, if all it takes to open a call center is throw some cash at the idea and hire a few experts, how long will it be before the rest of the world catches on, and throws some of their cash at this idea? Especially areas where the one difficult element — labor — comes a lot cheaper than it does in the US? Not long at all. This is already happening.

You could see it starting a few years ago here in the US, when the growth of call centers began to outstrip the available labor in their traditional areas, especially the midwest. Site selection moved south, west and north, and soon there were no areas in the US that *couldn't* make a case for why they should host a center. In fact, the locations that didn't compete were the ones that weren't competitive for other reasons, like government regulation (read: New York City). And places like Las Vegas, Phoenix, and Canada's New Brunswick became hot destinations.

Ireland sells itself as an English-speaking country with high unemployment and low labor costs, as well as an infrastructure designed to act as a distribution and fulfillment hub for the continent. Others in Europe are working just as hard to attract American call centers, with mixed results (again, usually depending on the openness and attractiveness of the business climate, less on technological factors).

And so it moves around the globe. This season there is a lot of attention being paid to Australia, which for now is a hot hub of call center activity. American service bureaus have been building centers and alliances down under. The attraction of that market is not as a conduit into the US, but because of its own wealth and domestic prospects. Here's an idea: As the economies of south Asia continue to implode because of currency and structural problems, will they not look to call centers as a fast way to put highly educated, well-skilled people to work, and build their own service sectors at the same time? If they can build a call center in a matter of weeks, staff it instantly, and do it on the cheap (with government subsidies), will they make a run at the Australian call center market, or the Japanese? I suspect it is only a matter of time before the commoditization we see here in the US is replicated every time you have an expensive, affluent call center market surrounded by an emerging, highly energized, technologically capable one. In addition to the Australian/south Asian combo, how about the European countries (some of them extremely dysfunctional in their call center industries and their eager competitors in the near-Mideast (Turkey? Israel?). Remember, most of these existing industries are still in their infancy compared to the size of the US market. British Telecom told me in 1995 that the number of call centers in Britain was less than 2,000. Even if that's doubled or tripled by now, it's still small compared with North America. And the UK has a powerful, very advanced industry.

And all we've talked about here is outsourcing. As outsourcing faces enormous price and competitive pressures, equipment vendors face unbelievable opportunity. Sell basic switches to call centers in the Philippines. Sell headsets to Turkey. Sell predictive dialers to Singapore. It's a world about to explode with call centers.

Chapter 25

Are Call Centers an Economic Indicator?

In these times of frenetic boom in the U.S. economy, it's only natural to look for signs that the end of the good times is close at hand. Business cycles may fade and elongate, but they never really go away — a downturn (or at least a significantly restrained upturn) will come.

The question is not whether, but when. And an even better question is: are there any signposts we can use to identify a change in national economic fortunes ahead of the curve?

In an informal way, the call center industry may be a leading indicator of the economy-to-come. It's informal for two reasons. First, because of the nuanced variety of components that make up the industry, and the many different ways you can segment it (horizontal vs. vertical; service vs. sales; equipment vs. software vs. services, etc.).

Second, there is the anecdotal nature of the business itself. Hard statistical data on the use and spread of call centers (and the industry's component equipment categories) is difficult to nail down. Much of it is hyperbolic anyway-optimistic projections of the adoption of new technologies put forward by a vendor community with a burning interest in having those forecasts come true.

To get a sense of where call centers fit into the larger economic picture, we need to have a sense of how companies are using their centers. Ten years ago, we would speak in terms of several clearly defined functions — inbound service; outbound telemarketing; and inbound direct response sales. These functions rarely co-existed within the same centers because they needed different technologies to operate properly, and used completely different business models for management and performance measurement.

Today, the shadings are much grayer. Now things co-exist everywhere. Call centers are the focal point for a wide-ranging discussion of the concept of "customer care," including such questions as the role of the Internet, electronic commerce and customer self-service in staying competitive and profitable. Companies see their call centers much more strategically than they used to. And it is rare to find an industry that ignores them completely (though some industries are generally slower on the uptake than others).

At the same time this has happened, the dominant force in call center technology has shifted from the traditional telecom companies to the application software providers. This shift, though much commented on, is profound. It caught many of the switch companies by surprise, forcing them to open their switches, to adopt standards, some of which was frankly hard for them to swallow.

When a call center manager considers a problem and looks for a solution, it is more likely that the solution will be found in one of two places: a software application or a change in his or her company's business practices.

What is the result of this? That hardware companies end up needing alliances with software app developers as much as the software companies need them. This is one key reason why you are seeing the wave of acquisitions these past few years — because developing new apps is expensive and redundant when there are small forceful companies out there already. Because the end-users of call center technology — the call centers themselves — are impatient with half-assed half-solutions, which is what a lot of vendors were feeding them in the early '90s.

In order to solve practical, everyday problems, and do it in a way that keeps the call center infrastructure competitive and strategic, and open to the new ways of doing business that are clearly coming, they need more integrated hardware/software combos that come from vertically integrated top-down vendors.

Software companies are buying vertically — Siebel buys Scopus to extend sales force automation with customer management. Aspect buys Voicetek to add IVR hardware and app gen strength to its switch. Quintus buys a consulting firm, and a CTI middleware developer. Davox buys a CTI company. Everywhere you look, it's easy to see that the food chain is tightening. It's clear that without relationships up and down the chain there's little a small company with good technology can do to thrive.

What this tells me is that the people who run call centers are more demanding about what they buy. They are skeptical of new technology that is not strongly integrated with existing systems and that does not show clear, demonstrable enhancements to productivity and profitability.

And what of the outsourcers? A recent study that projected major growth in the call center industry also cited major dissatisfaction with the outsourcing community. Outsourcers are having trouble making money. They are in a business with low barrier to entry. There is more call center capacity than is needed, at least in the U.S.

But the more strategic value a company places on its call center — indeed, on its relationship with its customers — the less likely it is to want to rely on an outside service provider to handle daily business. The Input study cited high turnover and low training at outsourcing centers. This does not bode well in a business that absorbs much of the high cost of running widely distributed centers with few of the soft-dollar benefits that internal call center operations reap.

If you take these two trends at the same time — the weakness of the outsourcers and the consolidation of strong software/hardware vendor combinations, I think you get a mixed picture of the call center industry as a whole.

The vendor community as a whole will continue to make money, but is being separated quickly into haves and have-nots, with the have-nots looking for partners. That is a game of musical chairs in which there are far fewer chairs than there used to be. And the music is getting faster.

The call center industry (if you can still call it that) is an industry in transition. It mimics the larger economy, which is also in profound transition. There are cross-currents, areas in which things are going dramatically well: extending corporate reach globally, for example, or using technology to promote personalized customer service.

There are ways in which things are not going so well — figuring out how to train and keep staff, or how to differentiate one service bureau from another.

Only an economist could tell for sure whether the industry leads the economy or the other way around. In any case, it's important that we try to see the relationship between the two. Call centers are in existence because companies have recognized the importance of customer contact. Economies are in existence for that very reason, too.

chapter 26

Telecom Merger Mania: Why It's Good For Call Centers

Once there were seven RBOCs. Once there were three major long distance carriers, and WorldCom wasn't one of them. AT&T combines with TCI.

I'm not going to deconstruct the reasons why telecom companies think bigger is better. These companies are jumping on the express train to consolidation, trying to make sure that no matter what pathway customers use to talk to one another (internet, telephone, cable modems, wireless, whatever), they have the pipeline into the customer's premise, and the money to lay more pipeline.

Much of the analysis of these mergers focuses on the financial benefits the companies themselves reap from combining, and the effect this will have on the general consumer or business customer.

But call centers are not general business customers. They are, if anything, the most specialized and savvy telecom consumers around, and they will be directly affected by a change in the basic telecom landscape.

If we stipulate that these mergers lead to a field of stronger companies (by no means certain), then call centers can look forward to a somewhat richer set of options than they have now (oddly, from fewer players). Stipulate also that regulatory issues will muddy everything, especially what I'm about to say about the combination of various kinds of networks.

Stronger companies will be able to put more advanced technology into the core telecom network. Already we have seen the beginnings of this, with customer-premise directed call routing features in the long distance networks. Transaction processing in the network is one of the best ways a carrier can augment their network to benefit call center customers. The majors already offer this, and the offerings are getting better. From their point of view, anything that adds to the telecom minutes used is a good thing. If they can get a call center to use a network-based front end like IVR, speech recognition or pre-call routing, and at the same time remove that capital cost from the call center, both sides win.

In addition, as they enhance the basic voice network with intelligent features, combinations of companies with different strengths bring different networks together; GTE and Bell Atlantic, for example, combines wired and wireless, local and long distance. AT&T and TCI, perhaps the most out-there combo yet, is gambling on the cable/telephone/internet combination. The best bets for call centers are going to be mergers that improve the quality of the core network to add features call centers can use: web-based transactions and voice calls, for example. Anything we can imagine that goes under the label "e-commerce" is a possibility.

Ultimately the assumption of the internet (or some component of it) into the carriers' voice networks will offer call centers more choices for routing each call. Customers are not going to start initiating voice calls through web pages in large numbers. But call centers will be able to offer toll-free telephony-based calls that use the internet for some legs of the connection. This will reduce the cost of the call. It also adds a layer of backup, because the number of possible routes a call can take is multiplied.

This will fundamentally change what we think of as Least Cost Routing; when the cost is this close to zero, other routing factors come into play. Call centers can be more widely dispersed, because the cost of calls that cross borders goes down. These are possibilities, of course, not realities. But I'm fairly certain that few call centers will implement internet telephony-based routing on any scale until it's a securely integrated part of a major carrier's offerings. And even then, the call center is going to want to be insulated from the internet component; that calls for even more enhanced network services, and a cycle is born.

Unlike traffic on the voice network, internet traffic isn't measured (for billing purposes) in minutes. Therefore, the only way to make money from building network traffic is to create revenue on every call, and take some of that revenue as part of delivering a service. (Like a routing front end, or even better, a transaction processing and fulfillment system.) That's why telephone/web combos will be the hottest offerings from carriers in the next five years. Because if they can take everything but the agent out of the call center, handle everything in the net and charge you for doing it, and keep your costs down by siphoning off 40% of your calls (like an IVR does), won't you go for it? If your two biggest costs in running a center are labor and telecom, and the telecom is willing to absorb both labor and capital equipment costs, are you going to say no to enhanced network services? Many will say yes. Enough to keep the carriers developing the services.

The connections between AT&T and British Telecom are interesting, too, because these two giants working together holds out the promise that true, cross-border carriers will be able to deliver international call processing as a global package. They will be vying for a piece of an international telecom services market estimated to grow from $36 billion this year to $180 billion in 2007 (That stat comes from the New York Times.)

AT&T is scuttling its other two existing international alliances, Unisource and WorldPartners. That's going to leave a gaping hole in the marketing of combined services, particularly in Europe, and it's going to leave some of the national carriers in the smaller call centering countries without a partner to help smooth the transition for US companies locating centers in that region.

The call center industry will be watching the AT&T/BT alliance to see if it really becomes a branded, international carrier that can seamlessly connect centers in the US to pan-European centers, and watching to see what effect it has on the regulatory climate in Europe and other areas.

The merger frenzy will ultimately lead to a carrier that functions as a full service backup to a call center: everything from the telecom minutes, network services, as well as outsourcing for overflow and backup, for corporate messaging traffic (like voice mail, e-mail, internet and faxes), and even systems integration services. They are well suited to the umbrella role: companies like US West, Ameritech, and the larger long distance carriers work with equipment and software vendors in many call center categories to offer turnkey call center packages: everything you need under one roof, for a 10-15% price premium over doing it yourself. For that premium, the carrier will certify that everything works together, and often provides one-call multivendor technical support. This trend will accelerate.

There is a price to be paid, of course. It's hard to project one or two or five years down the line and make any assumptions about what telecom costs will be like. They may be higher, because you'll be paying for more than the bare minutes. They may be lower, because the internet may reduce call costs to nothing, leaving the carriers to provide them "almost-free" and charge for the value-add. It's a cloudy, unpredictable world. All we know is that the topsy turvy corporate shuffling that's going on will result in some strange new choices, some exciting new technologies, and a lot of options for innovative call centers.

chapter 27

Sweatshops for the 21st Century?

There was a surprising article published by the BBC to their website in 1998, entitled "Call centres: the new sweatshops of Britain?"

The article was cursory, but had several interesting things to say. First, it cited statistics, which since they are hard to come by, I will repeat here:

- There are 7,000 call centers in Britain, employing 200,000 workers.

- One in every hundred workers in Britain works in a call center.

- And most workers' tenure is less than two years, due they say to the stress of monitoring and surveillance. (These stats are sourced back to a London School of Economics study).

I'll leave the monitoring debate to others. What's interesting at this point though is that Britain is wrestling with success—and what that success means for the kind of economy the country wants to have.

Sweatshops for the 21st Century?

The appellation sweatshop is, I think, unfair. Call centers are, it is true, an industry that employs those who are minimally skilled and nearer the bottom of the economic ladder than the top.

However, sweatshop brings with it a connotation of exploitation and poor working conditions, along with a sense that the job you are in is a terminal one — the end of the line for someone with nowhere to go and no future prospects.

This is not the case. Call centers are a job creation engine in small and medium sized communities. They demand huge numbers of workers, and can transform a community in a short time. They absorb excess labor in a marginal economy.

That's why the competition for locating call centers has been so fierce over the years. And especially so in the last five years, as the factors that kept call centers out of certain areas were mitigated.

Originally, way back when, the most important element of where to place a center was the availability of strong telecom services. Through the 1980s, this most often meant that you were close to an AT&T point of presence.

Over time, robust telecom became so widely available that the selection focus shifted to a more reasonable assessment of relative cost, particularly with labor. To run a call center of any size for any length of time, you are going to need a labor pool deep enough to be able to replace the one you start with several times.

Burnout is a real factor, no mistake. People get tired of dealing with phones all day, of dealing with irate customers, of the constant repetitive process of working in a call center. That's one of the reasons why people leave and the general turnover rate is so high. (One US-based consultant recently told me that when he recruits for large centers, he looks for three times as many workers as he thinks he's going to need.)

Creating Skilled Workers

But people who leave a call center are not thrown back to the bottom of the workforce. Remember, one cause of high turnover is the deliberate selection of a transient workforce in the first place — people who are flexible enough to work odd shift are often students, women with small children, spouses of military personnel, the elderly — all groups of people that tend to move around.

High turnover among centers that hire from these groups does not necessarily reflect a stress in the workplace.

People who leave call center work often come away with a skill base that's incredibly valuable in other kinds of work. Computer skills, for one, but also very valuable interpersonal skills, including the ability to focus on customers, to solve problems and to prioritize. Call centers create articulate, ambitious, economically viable workers. And that's why small communities fall over one another to attract centers. They create a tax base of skilled workers.

Why Governments Want Your Call Center

From the call center point of view, the key element that distinguishes one potential site from another is cost, rather than telecom. It's the direct cost of labor, and the availability of a pool of continuous hires, but also indirect costs as well.

These come in the form of taxes (and incentives, which are really government attempts to mitigate costs by leveling out the playing field). Local development authorities will offer things like real estate tax abatements, carrots like money for training, for each job created, etc.

Where it really gets interesting, and where call centering companies ought to pay the most attention, is where communities get local groups into the act beyond the government. If, for example, the locality is interested enough to convince an educational institution to start a training or certification program, that's a strong indication that you'll be able to staff for the second and third turnover cycles.

In the US, this has played out over the last ten years in an incredible decentralization of call center locations. They are in the biggest of cities (Los Angeles, Phoenix, Las Vegas) and the smallest towns in rural areas like Oklahoma, Alabama, the Dakotas.

And for all practical purposes, the border between the US and Canada doesn't exist, as Manitoba, Nova Scotia and other northern areas compete as effectively for American call center business as any US-based region.

In the US, where you locate often has more to do with the peculiarities of your own business than any call center qualifications. Often the new center has to be near an existing center, or near a corporate office, fulfillment center, or factory, near a client or the CEO's home town. (Yes, that happens.)

Outside the US, there are more factors, like the kind of product or service you are offering and the marketing that you do to support that. In Europe, for example, the notion of the pan-European center has strongly taken hold because in a fragmented market, many think that the best way to reach a market that's multilingual and cross-border is with larger, diverse centers. The Netherlands and Ireland have historically done well attracting US companies that are trying to reach all of Europe in the most economical way.

Other countries have tried to take advantage of the size of their own domestic consumer economies to create home-grown call center industries that can then be exported; the UK is the leader at this, with the second largest call center industry in the world (or third, depending on how you measure it).

I understand the ambivalence felt by the British when they suddenly realize how fast and far this industry has grown. More people work in call centers in Britain than in the coal, steel and automobile industries combined, says the London School of Economics. The industry in Britain, already mature, is now facing some of the same structural human issues that we are seeing in the US — involving things like privacy, worker motivation, and what happens when different areas compete for call centers.

These are healthy signs, and more so because they represent the UK's internal industry, not one that looks across borders for success.

Other smaller European areas that thrived from attracting external call center business (i.e., call centers that took or made calls to or from other countries, like pan-European centers) face strong competition from newly awakened giants like France and Germany.

These larger economies, with little or no domestic call center industries to speak of, are focusing on attracting outside centers, and in doing so might make it difficult for countries like Belgium or the Netherlands to continue to attract centers.

Again, it's all because governments are alerted to the benefits of centers — job creation, mainly — that they are trying to attract centers. They are clean, nonpolluting, skilled jobs that come at a low cost to the environment or local economy. If these are the sweatshops of the new Britain (or America, for that matter), we should count ourselves lucky.

chapter 28

Building An Industry From Scratch

We take it for granted that a call center can locate anywhere, but is this really true? Is it possible for a small country, starting almost from scratch, to build itself a call center industry?

That's the question that El Salvador is grappling with right now, as they try to attract outside call centers as a way of lifting their economy.

The possibilities are tantalizing. On the plus side, they offer a Spanish-speaking workforce to answer calls from the US or Latin America. But there are high technological, human and business hurdles that stand between them and this clean, efficient industry.

In the last issue, we grappled with the idea of why call centers are attractive to local development authorities. It seems to boil down to jobs: call centers provide plenty of them, and they're not bad jobs, as jobs go.

Despite the tax incentives a locality has to dole out, attracting centers boosts a tax base through more skilled workers, and the ricochet of their spending through a local economy.

It's also commonly held that you can put a call center pretty much anywhere that it makes business sense for you-that you don't have to worry anymore about where there is good telecom or where costs are low, because for competitive reasons, costs are low all over.

That kind of thinking has led in the US to fierce competition among smaller venues, often down to the county level, for even small- and medium-sized centers.

Outside the US, it's led countries like El Salvador to an interesting crossroads: does it make more sense to develop a domestic call center industry, or should they try to become outsourcers to a larger market, like the US? Or a mix of the two?

I recently sat down with a Gerardo Tablas from El Salvador's development agency for a conversation about the prospects and hurdles that country faces as it tries to create a more robust economy for itself.

A little background: El Salvador is a country of six million in Latin America. About one-sixth of that population is actually in the US, making for an interestingly skilled workforce, and one that has more experience with call centers (from the consumer side) that you might find in other countries.

El Salvador is geographically close to the US, particularly to the Spanish-speaking southwest and to Florida.

The country has several mid-sized call centers already, mainly in banking and for the national airline's reservation system. (A million Salvadorans in the US generate a lot of phone calls for flight reservations.)

They have been making the case for some time that they are a desirable location for an American outsourcing company to open (or take over) a center that would answer calls from either the US itself, or from the wider Latin American region.

The idea behind that is the same as building a shopping mall – you want to have a big name as the anchor tenant, so others will feel comfortable moving in. They've been negotiating with a couple of the big names in outsourcing, but so far nothing has been signed.

Gerardo Tablas argues that an American company could come in and purchase one of the country's existing centers and, using their expertise, wring greater efficiency out of it than the local operators.

This is the traditional outsourcing argument, that a company that's not in the core business of running a call center can benefit from having an outsourcer come in and run the center.

After that "showcase" center is operating, attention would turn to outsourcing some of the capacity to answering diverted calls - those that wouldn't terminate in El Salvador normally. Those calls could be American calls answered in English or Spanish, or Latin American calls in Spanish.

There is no reason, for example, that calls from consumers in Mexico or Venezuela can't be answered in El Salvador. As long as it's cheaper, and just as reliable, the original argument holds: you can place your call center anywhere.

There is a certain logic to this argument. Certainly, the prospect of call center jobs means real improvement in the economy of El Salvador. But there are pitfalls.

The Downside of An External Industry

The problem is, when you rely on an external industry to light a fire under your economy, you have to hope that the industry remains vibrant and expansive. The American call center outsourcers are a thin rope on which to hang those hopes.

Outsourcers often rely too heavily on just a few large clients; when stresses force those clients to cut back, the outsourcer is left with revenue holes. Plugging those holes sometimes means reducing costs, which lately has meant closing or consolidating centers.

Since a country as small as El Salvador can probably support a very few large centers of the kind that an outsourcer would need, they risk putting all their eggs in one basket should external factors force the outsourcer to backpedal and reduce or shut down their Salvadoran presence.

Also, an externally-focused industry is an industry at the mercy of the world and regional economies. Should Latin America follow Asia into recession, now or five years from now, the gains made through the development of call centers could be erased when consumer contractions reduce the number of calls that need to be answered from Mexico or Miami.

I wouldn't advise small countries against trying to attract outsourcers - despite the pitfalls, it's the fastest way to get a strong boost from the call center economy-I would recommend making a strong effort to build a local, internal call center industry. Any country advanced enough to be considered as a site for an external call center by definition has enough of a consumer economy to begin supporting local call centers.

It starts with banks and airlines, telephone companies, credit card issuers, catalog retailers and direct response sales. In many cases, small centers already exist for some local businesses. What a small country (or region) should do is facilitate the spread of knowledge about the industry as it already exists in more developed areas like the US and Europe. They should form a government/industry council to share information about operating centers within companies.

Encouraging local companies to do business by phone raises the bar of what's expected, no less in a place like El Salvador than in Kansas or Dublin or Los Angeles.

And in the early stages, it is absolutely essential that a small country with serious intentions-whether internally or externally focused-needs to develop an educational training program to support the industry. El Salvador, with its six million, needs to be able to convince anyone thinking about a call center there that they will be able to hire staff, and keep on hiring staff long after the first year.

They should begin a crash course to train people who can speak articulately, in as many languages as possible, with as much sensitivity to customer needs as possible. They should incorporate computer training into the secondary and vocational school systems, whether or not they are successful at attracting centers in the early stages.

This hidden cost will pay dividends down the road. The country will then be better off in the future, when call centers, merged with the internet, become more complex centers of voice and data and transaction processing.

The El Salvador experience is not unique. Countries all over the world are considering the benefits of call centers.

From the Philippines to Belgium, locations are weighing their attractiveness to this cross-border industry in the hopes of boosting the fortunes of their workers and creating a solid, high-tech economy.

Whether or not El Salvador succeeds in attracting or building a call center industry, they are grappling with complex issues that have no easy solution. Only with time will we know if they come too late to the party.

chapter 29

Telemarketing: More Than A Phone Call

What happened to old-fashioned telemarketing? It has come a long way beyond annoying calls at the dinner hour. It has become a broad-based, international industry that employs millions and garners billions in annual revenue. And yet many who come to telemarketing from some of the similar disciplines (like the direct mail business, or database marketing, to name two) still see the industry the old way.

It might surprise many to know that telemarketing today is primarily an inbound business — that the telephone is America's primary way of doing business, and that in an increasingly global economy, we are its leading practitioners. That the rest of the world is rushing to implement many of the tools and techniques we have pioneered. Telemarketing is the natural outgrowth of a trend that goes back as far as the original Sears catalog: the easier a business makes buying, the more its customers will buy.

In the 1990s, telemarketing has become completely entwined with customer service and with direct response TV. Customer interactions have so many flavors now that there is no room for lost opportunities to sell, cross-sell or up-sell.

Thanks to telemarketing, there is a broad based consensus among companies that every time a customer and a company come together, no matter who initiated the contact, there is an opportunity for relationship building. And an opportunity for a sale. This is due to the trail blazed by the telemarketing industry.

Telemarketing reaches out to the customer in an attempt to inform, entice, and sell. But raw outbound telemarketing is merely the tip of the iceberg. Real, professional telemarketing is a subtle, nuanced practice that aims to connect to a potential customer. Technology enables a telemarketer to know with strong certainty that the person he is reaching out to is a real prospect, that his product has value to this particular potential customer.

The Era Of The Engaged Consumer

To a greater degree than ever before, consumers are partners in their own marketing. This is not just a question of technological savvy — they are also concerned about issues like price, quality and convenience. There are very real reasons why people like to go to malls, shop from catalogs, order from Web pages and get cash from ATMs. These introduce efficiencies and small pleasures into the purchase.

Customers are willing to meet companies halfway, willing even to let them into their homes in the form of infomercials and unsolicited telephone calls, because they know (sometimes unconsciously) that these tools offer them convenience and lower prices.

They know that the telemarketing sales call is only half the equation — that this is the same industry that staffs centers with people who walk them through customer service problems on a Saturday night. They know that they can buy books or CDs or have flowers delivered to someone across the country in just a few minutes with a credit card and a telephone. Small pleasures, made possible by billions of dollars in technology and decades of accumulated experience in making it easier for people to buy what they want to buy.

In recent years, the technology has accelerated the trend toward engaging the consumer. In just the last 24 months, Web-enabled call centers have come on-line, offering customers the chance to gather as much information as they want or need before they pick up the phone to place an order.

Old-style outbound telemarketing still exists, of course, but it is so well targeted, so specialized, that it is now just a subset of a larger and more complex set of interactions.

Combined media now do the job of the telephone — if you want to reach a group of likely buyers you are now more likely to use a combination of media, ranging from e-mail to fax to direct mail to a telephone call to direct response broadcast media.

The power of these media in combination is awesome. And the control that individuals have over the gathering of their own information is similarly awesome. The perception of telemarketing — that it exists to sell something to somebody who is ill-informed — lags far behind the reality. Reality is that even if you don't call them, they will come banging on your doors to find out what you have for sale.

Telemarketing In A Multi-Channel World

When there are so many ways to reach customers, and so many reciprocal ways for the customer to talk back, it can often be hard to discern exactly where the line is drawn between mass marketing and individual, personalized sales.

In fact, you can make a pretty good case that we are entering a period wherein every consumer will be known so thoroughly by his or her past preferences that the old notion of telemarketing, with its connotations of scattershot randomness, will be replaced by a highly targeted and scientific "mass personalization."

Database marketing brings so much to the table that improves telemarketing. It brings the technological sophistication that aims to screen out of any campaign those for whom the message you deliver is wasted. There are very good reasons why companies are turning to things like data warehousing and mining — because hidden in those mountains of corporate customer data are patterns of who the customers are and what they really want. It rarely makes sense to throw wideband messages out at high cost for little return. Database marketing, in all its forms, has as its mission the narrowing of the focus. In the past few years the industry has been given tools that amplify its simultaneous need for more focus and higher volumes: the predictive dialer; high performance PCs; demographics and psychographics; and the multi-use inbound and outbound call center.

Telemarketing will be (and is already becoming) the most effective way to create consumer awareness of things they are known to be interested in. And of doing it in a way that's so cost-effective that it keeps prices low.

Just as telemarketing grew in the 1990s to encompass inbound sales and direct response, in the coming decade I am sure we will be using that term to include things like personalized TV and Web advertising, direct fax, direct e-mail and other things we can only speculate about now.

Part Six

Building the Call Center of the Future

IDENTIFY the customer.
PRIORITIZE his needs.
Bring all the information
resources of the company
to bear on that intersection.

chapter 30

The New Role of the Call Center

In recent years, technology has remade the call center — it has given us better tools for analysis, forecasting, and for real time operation of centers. With new tools at their disposal, managers are capable of a lot more. So, it bears asking: has the essential mission of a call center within an organization changed as well?

Not entirely. Instead, I think there has been a reassessment of the need for coordination between center and enterprise. In many cases this is less of a "re"-assessment and more of a first-time look. A number of forces have converged very recently, and have given companies that operate call centers a dramatic opportunity.

1. CTI products appeared in the marketplace faster than the call center industry expected. The results from call centers that installed CTI in the last few years (as early adopters) show good productivity results from the technology, thanks to the efficiencies of screen pop and intelligent call routing.

But CTI is expensive, and introduces a lot of complexity to the day-to-day management of a center. So the decision to apply a layer of computer telephony intelligence to a call center is often made in conjunction with other departments, especially MIS and financial management.

2. The growth of the Internet. Actually, despite all the hoopla over the Web and the Internet, from the call center perspective, this is just one thread in a tapestry of integrations: web, IVR, fax, video, e-mail; they all function as alternate pathways into a call center. In the aggregate, they change the dynamic of how a center answers calls (assigning different priorities to the different methods, for example) and force a center to come up with a new set of metrics to benchmark performance.

3. Open standards — and not just the CTI standards that help developers. Things like Java, HTML and COM/DCOM make a big difference in the way they allow a company to share data among employees, whether or not those employees work in the call center. For the first time, we live in a world where a call center in San Jose can pump data out to an analyst working in London, who can fire off a product report to New York — with all the real-time data intact.

What this means for call centers is that their role and prominence rises within an organization — at the same time as their hold over the data they generate diminishes. I think that the data that call centers spit out (mostly ACD data) has a lot of application to other employees when formatted intelligently. The more this is done, the more the MIS department will look on that data as falling under their purview. Maybe this is a good thing; only time will tell.

While call center management may have to cede some control over their information infrastructures, they will take on more of a customer management role. The job of the call center will be to synthesize all the existing pieces of information about customers-from ongoing interactions and from legacy databases-and address the needs of each customer on an individualized basis. I recently saw a talk given by consultant Martha Rogers, wherein she described it this way: "Treat different customers differently, but treat each one consistently." (She was talking about a company's strategy in general, but I think the call center is where this approach will be put into practice.)

By the year 2001, the role of the call center will be to "mass customize" service: to identify the customer, prioritize that person, and bring all the information resources of the company to bear on that interaction.

The result will be faster service, a higher customer retention rate, and (believe it or not) lower costs.

chapter 31

Customizing Every Interaction: What All This Technology Is Really Good For

Several years ago, I tried out a new way of describing a call center. It's not a physical place, went the argument. It's a set of functions that can be carried out in any number of ways, in lots of configurations, many of them virtual. It's customer contact, no matter how that contact is handled. (I was trying to get a grip on the way media blending and the Internet were changing the way call centers were managed.)

Now, we are at a transitional moment in the call center industry. Technology has leaped ahead, and all the tools that we could possibly want or expect (or assimilate) are available. A really motivated company with lots of money to spend could put together the kind of integrated call center/customer service/database system we dreamed about a few years ago.

Call centering has also ceased to be an arcane specialty. The people running these centers are as likely to be networking experts as they are telecom people. (You can, in fact, make a pretty powerful argument that call centers shouldn't be thought of as part of the telecom industry at all.) What it takes to run a modern, top-notch call center requires more in the way of people skills and organizational savvy than the ability to program a switch.

In fact, they don't want to waste their time knowing what wire goes in what socket. That's not interesting, and it's not helpful in achieving the core goals of the call center. Those goals:

- Create as many points of interaction between company and customer.

- Minimize the cost of those interactions.

- Identify those non-customers who might become customers.

- Give everyone outside the company the tools to gather information about the company.

- Gather as much information about those customers as possible.

- Provide that information to anyone in the company who can use it.

The key words here are company and customer. Stop and think about that: the only thing that matters is bringing together the people with the money and the people with the product. The call center is the best tool ever invented to do that. What is not interesting any more is the call. On a low level of operation, you certainly want to measure calls, because that's the metric of productivity in a center — how many calls did these agents answer in how short a period of time, equals how much it cost you to run the center.

In the future, no one will care much about that. What everyone outside the call center is measuring is customers, and the results of the interactions. The call center's role is shifting — from a place where calls get answered to the place where information is exchanged.

Precious few call centers have made this transition yet. So many are still operating according to the old rules that it may seem premature to talk about this radical shift. But the fact that the technology is out there renders the point moot — this transition will happen, is already happening. If CTI is only used in 2% of call centers (a low estimate), and there are 100,000 call centers in North America, there are still an awful lot of state-of-the-art centers out there testing new ways of operating in this data-centric environment. And you know where those CTI-enabled centers are: hidden away in plain sight — you touch them every time you call Schwab and speak the name of the stock quote you want; every time you track a package with FedEx's web site, whenever an airline calls you to tell you your flight will be delayed. At America's top companies, the ones that have the most riding on every customer interaction.

In the ideal world, every customer will be handled as if she is the company's only customer. Every interaction she has with the company will be enlightened by all the relevant information about her needs and desires. All the data she has ever expressed about preferences and background will be saved, and brought to bear on the interaction intelligently. It is not simply a matter of collecting data and storing it in a backoffice database somewhere, where it is reported on and forgotten. It must travel in two directions. When a company knows something about a customer, it must use that knowledge to build the relationship between the two parties. And it must do so relentlessly, so that the customer reaches the point where she is doing business with the company because she has so much invested in the company that it's simply easier than switching to a competitor.

And when you do that for every single customer interaction, across all customers, whether it comes in the form of a phone call, a web visit, an outbound telemarketing call or an in-store visit — that is the mass customization of service.

There are really four ways of practicing customer service. The first, most basic method, is to treat every call just as a ringing phone — somebody must answer it at some point. When lots of phones ring, that means trouble. This is the triage stage.

Then there is the traditional, early '90s method of call-centering-as-usual. Those phones are ringing? Get an ACD to route them, separate the agents into groups, and by God we can handle the volume. The goal of this method is cost-containment.

Then it gets interesting. There is a natural progression from triage to cost-containment. From there, companies realize that the call center offers an unprecedented opportunity to gather information, and use that data in a rudimentary way to get new customers, or sell existing customers more product. This is the old CTI paradigm: "Oh, I see you've bought a green shirt in the past, Mr. Dawson. We have a new line of green shirts, would you like to try one on sale?" Crude, but effective. What differentiates this third stage of service operation from the cost-containment stage is the more optimistic view of the call center as a corporate asset, rather than a cost-center.

Taken to its logical extreme, using current technology, is the stage four call center: where every interaction is completely customized from the ground up. Service is designed as a seamlessly integrated component of the corporate-wide customer retention strategy. It's reflected in the kind of data gathered and the information that makes its way to the agent's screen. This is starting to percolate out into the public consciousness: look at the success of Levi's jeans that are computer designed to fit women; or web sites that let the customer choose how to view them (Yahoo, for example). Once the processes are designed, the cost per interaction is negligible. The revenue opportunities are enormous.

And yet, most companies today are stranded somewhere between stages two and three, between containing costs and starting to think about ways to add intelligence to their calls. Until they stop thinking in terms of calls and start thinking in terms of customers, that's where they will stay.

chapter 32

The Call Center, 2005

The call center of the future will look an awful lot like the call center of the present. There will still be people sitting in cubicles, talking on headsets, working on computers. The look and feel of the center, the operational techniques, will remain relatively stable.

The differences will be below the surface, in the kinds of things that center is doing. There will be subtle changes in some of the tools, and in the way the call center is linked to the rest of the organization. Let's look at the components of the call center, from the ground up.

I don't believe that five years out call centers will be using any dramatically different technology for answering and routing calls. The ACD will remain the dominant piece of call center hardware; it will of course be smaller, more open, and augmented by stronger software to enable more intelligence to be applied to each call. Rudimentary ACD features will be standard on business phone systems, making every company equipped from Day One to run some kind of basic departmental call center.

Call centers will still be buying their telecom services from the same three companies they buy from now. It's not practical to predict rates much lower than what call centers get now as huge volume buyers.

However, it is reasonable to assume that five years from now, international expansion of linked call centers will be much more common, requiring a lot more international overflow, and a lot more data travelling along with each call. Expect that rates between countries will fall, partly to encourage the growth of new centers in hungry countries.

There are many countries in Europe, for example, that gnash their teeth helplessly as some of their neighbors attract huge investments from US call center companies, while they get little or nothing themselves. The business climate will change, starting with the overseas carriers. If not, the call center job creation engine will land somewhere else.

Some wildcard possibilities for popular call center locations in 2005. International: Israel and the rest of the mideast; the Philippines; China; Mexico; the Caribbean. Domestic: West Texas (in and around San Antonio); the upper Midwest; and (believe it or not) New York City (where economic development efforts and enterprise zones may combine to offer attractive enough incentives for centers to locate there).

In 2005, the agent desktop will remain PC-based, augmented by powerful intranet browsers. On the outbound side, nothing will change. Calls will be downloaded to the agent from the list processor just as they are today. The predictive dialing engine will either be a fully integrated component of the switch, or an application residing on the network. Both will be available, but the era of the closed, proprietary, standalone dialer will be coming to a close.

Inbound calls will be screened by the switch through advanced computer telephony applications, and the calls will be served to the agent only after a tremendous amount of data has been sifted and analyzed.

First, data about the call will be piped out of the call center into the MIS infrastructure; while the backend systems categorize the caller and assemble the information the agent needs to deal with that person, the switch is using the call data to play the appropriate waiting message, put the call into the right queue, or query a network of linked centers to find the most efficient place to handle that call.

By 2005, much of the work that needs to be done in connecting call centers to the backend databases will (hopefully) have been done, and will arrive as working off-the-shelf products (rather than as the clunky custom solutions that are so expensive and so unwieldy today).

Managers will have better tools at their disposal for measuring agent productivity — and that productivity will be seen more in terms of revenue produced and problems solved than calls handled.

In 2005, the competition for call center labor will be so tight that a great deal more attention will be placed on proper training, motivation and agent retention programs. And still, turnover will be high and call centering will a career path in name only. By 2005, the call center industry in the US will still not have an effective trade association, though I think by that time we will have the beginning of a core of training and certification standards for reps and center managers.

Despite the raft of technology that purports to "empower" agents by reducing the repetitive, annoying calls and increasing the amount of "knowledge work" they do, call centers will still be staffed by an overworked, transient staff with little chance of promotion to a career track.

In 2005, the Internet's chief role as far as call centers are concerned will be to serve as the gatekeeper: the place where people go to get the core information they need to successfully interact with a company. Like IVR, the Internet will be the place where they check their account balance, get literature, view a catalog, etc. Then they will pick up the phone and call a toll-free number. The telephone is the most useful instrument yet devised to communicate by voice at a distance. The Internet is not now such a device, nor will it be by 2005.

It excels as a vehicle for self-service and data collection. And its intranet cousin will make a big difference in communication within a company. But anyone who thinks consumers will choose the Internet over the phone to complain about their credit card bill is living in a technological dream land.

There will be a lot of small, incremental changes, but generally call centers are slow to adopt new technologies. The slow progress of computer telephony is a great example. Call center managers (and the folks above them who hold the purse strings) don't like to commit to something that doesn't show an immediate benefit. In the next five years, the growth of call centers as an industry will be steady across borders, and the call center itself will be more tightly linked to the rest of the organization.

There will be turf battles within every company over control of the data generated by the call center, and often the call center will lose — when the call center takes on the role of data clearinghouse (as seems likely), MIS, as custodians of the existing data infrastructure, will have a lot to say about that data. The stature of the call center may diminish within some organizations, even as its importance grows.

This is not a pessimistic view. Realistically, things don't change all that dramatically over a five year time span. Unfortunately, technology being what it is, ten years is too long a span to predict. Tell me what you think — if any of this looks wrong, or if you have a different picture of the future (or the present), write me at keith@callcenternews.com. I promise to write back.

Part Seven

Technology

**TOOLS for finding
what you need — from
SWITCHES to SOFTWARE
and everything in between**

call center resources

This listing — by no means complete — includes companies that sell equipment, software and services into call centers. It's as current as we could make it, given how fast things change in high tech industries. If you have changes, find errors, or want more information, visit the electronic version of this list at www.CallCenterNews.com. That's the website of the *Call Center News Service.* You can update your own listing there, as well.

800 Direct
(818) 713-1092 (800) 888-5524
Fax: (818) 713-8426
www.800direct.com
800 Direct is a full-service agency, providing one source for telemarketing, including fulfillment, database management, inbound and outbound telemarketing and interactive services.

800 Support
(503) 684-2826 (800) 777-9608
Fax: (503) 639-3946
www.800support.com
General technical support outsourcing for hardware manufacturers, software publishers, ISPs, and major corporations.

AAA Best Mailing Lists
(800) 692-2378
Fax: (602) 745-3800

ABConsulting
(408) 243-2234
Fax: (408) 243-2236

Consulting services on fax, fax-on-demand and other fax/call center issues.

Aberdeen Group
(617) 723-7890
Fax: (617) 723-7897
www.aberdeen.com

ABI Companies
(813) 289-8808
Fax: (813) 289-6628
www.abiinc.com
Construction, design and real estate services for call centers. They have built a number of large and important centers, mainly in the Southeast US.

Acacia Teleservices International
(541) 484-5545 (800) 225-4151
Fax: (541) 465-9406
www.acaciatel.com

Accelerated Payment Systems
(800) 688-2230
Fax: (410) 584-9548
www.ncms.com
Automated check debiting system for call center transaction processing. (That is, callers can pay by check and the transaction is automatically verified, much like a standard credit card transaction.)

ACI Telecentrics
(612) 928-4700
Fax: (612) 928-4701
ACI Telecentrics is a call center outsourcer, employing more than 800 telemarketing reps for telephone-based sales and marketing services primarily to the publishing, financial, insurance, and telecommunications service industries. They operate call centers in seven cities: Twin Valley, MN; Devils Lake and Valley City, ND; Redfield and Pierre, SD; Lombard, IL; and Merrillville, IN.

ACS Wireless
(408) 461-3270 (800) 995-5500
Fax: (408) 438-2745
www.acs.com
Headsets, wireless and corded.

Active Voice

(206) 441-4700
Fax: (206) 441-4784
www.activevoice.com
Repartee voice processing and unified messaging applications. It's actually one of the most popular voice mail systems around, enhanced all the time. (They've popularized the "1 for yes, 2 for no" interface.) Also important is the TeLANophy system, which is essentially a LAN/voice mail interface— all your calls are viewable and listenable through your PC. I've tried it, it works.

Acuity
(512) 425-2200 (888) 242-8669
www.acuity.com

Aculab
+44 (0) 1908 273800
Fax: +44 (0) 1908 273801
www.aculab.com
Computer telephony hardware, particularly boards.

Acuvoice
(408) 289-1661
Fax: (408) 289-1201
www.acuvoice.com
Speech synthesis system and text-to-speech system. They've got an expert system that takes text and parses it into voice sounds. They point out that it delivers speech in natural sounding tones, not robot-speech or the typical "drunken Swede" sounds (apologies to my Swedish friends).

Acxiom
(501) 336-1000 800-9ACXIOM
Fax: (501) 336-3902
www.acxiom.com
Service bureau offering a wide variety of list management and data management tools. They are active in direct marketing, data warehousing and certain vertical applications of those technologies.

Adante
(760) 431-6400 (888) 827-5557
www.adante.com
Adante, which is essentially "ACD e-mail" as they described it to me. Call center agents use their web browser to connect to an Adante server (Win 95 or NT) which then gets them the "next e-mail in queue". It gives you all the same kinds of routing criteria for e-

mail that you get with a voice-based ACD, including behind-the-scenes integration with customer information. Interesting.

Aditi Corporation

(425) 828-9587 Fax: (425) 897-2900

www.aditi.com

Software for transacting customer service over the internet.

Advanced Recognition Technology

(408) 973-9686

Fax: (408) 973-9687

www.artcomp.com

Smart Speak voice recognition software.

Advantage kbs

(732) 287-2236

Fax: (732) 287-3193

www.akbs.com

Problem resolution software and customer support applications. The suite of products is called The IQSupport Application Suite.

Aegis Communications Group

(214) 361-9870

Fax: (214) 361-9874

Aegis provides outsourced telecommunications-based marketing, customer service and call center management services in the US.

Affinitec Call Center Systems/AAC

(714) 756-2700

Fax: (714) 851-6286

www.aaccorp.com

Call center management software, readerboard drivers, call accounting systems. Their products include TimeManager, ForecastManager, ScheduleManager and AgentView. Readerboard systems typically include both the display board and the ACD data extraction and formatting tool that it needs. Hence, you can often use them to extract data to a PC, in addition to (or instead of) the readerboard.

Ahern Communications

(800) 451-3280

Fax: (617) 328-9070

www.aherncorp.com

A headset distributor that resells lots of brands, including Plantronics and Uniden cordless, as well as speakerphones, videoconferencing equipment and a variety of spare parts for headsets.

Alert Communications

(213) 254-7171 (800) 333-7772

Fax: (213) 254-6802

www.alertcom.com

Service bureau and outsourcer of call center services. Alert operates four call centers in the Southern California area. These centers process both live operator and automated call processing applications, including Some of Alert's areas of specialization include the catalog order processing, software and hardware customer support, dealer location referral services, and Internet call response.

Allen Systems Group

(941) 435-2200 (800) 932-5536

Fax: (941) 263-3692

Alltel

(800) 440-3262 (800) 440-3262

Fax: (502) 220-6880

www.alltel.com

Alltel offers comprehensive call center solutions including consulting, implementation and dedicated or shared outsurcing.

Alpha Technologies

800-32-ALPHA

Fax: (360) 671-4936

Power protection, CFR Series UPS. ALCI industrial line conditioners.

AltiGen Communications

888-ALTIGEN

Fax: (510) 252-9738

www.altigen.com

AltiGen provides computer, telephony and internet solutions for small to mid-sized businesses. The AltiServ phone system uses AltiWare software to provide informal call center functionality in a single Windows NT Server-based telecom platform that also includes PBX, voice mail, e-mail, auto attendant and internet integration.

Amcom Software

(612) 946-7707 (800) 852-8935

Fax: (612) 946-7700

www.amcomsoft.com

CTI Smart Center: a suite of call center applications including auto-attendant, voice response and various messaging features. All of Amcom's products use ORACLE as the RDBMS system and use Windows 95 as the GUI user interface. The next revision will use Windows NT.

Amend Group
(214) 696-6900

www.amend.com

Site selection assistance as well as commercial real estate services for call centers. They do research on labor costs and economic incentives.

American Power Conversion
(800) 788-2208

Fax: (401) 788-2712

www.apcc.com

UPS, power protection & surge protectors. Power products can protect telecom, computer and networking systems. They also sell a series of desktop-based software that helps alert you to and manage power fluctuations.

American Productivity & Quality Center
(800) 776-9676

www.aqpc.org

Organization that does benchmarking and other call center-related research.

American Telemarketing Association
(818) 766-5324 (800) 441-3335

Fax: (818) 766-8168

www.ataconnect.org

Industry association for inbound and outbound telemarketers. Provides information, lobbies Congress, and has chapters around the US.

Ameritech
(312) 364-4300

Fax: (312) 849-3611

www.ameritech.com

Telco that offers a branded, turnkey, "end-to-end" call center system to their customers. Basically, they work with a menu of hardware and software providers to cobble together call centers that are certified to have all the integrations work. This is the latest wrinkle in up-market systems integration. Ameritech's roster of partners includes Nortel and Rockwell for switches, APAC for call center services, TCS and Mosaix for applications and Manpower (one of the world's largest staffing firms). Working closely as well is IMC, a consulting firm that specializes in optimizing call center performance.

Amtelco

(608) 838-4194 (800) 356-9224

Fax: (608) 838-8367

www.amtelcom.com

Both developer and more turnkey call center systems, including 1Call, which offers modular ACD functions, directory systems, departmental registry, and e-mail.

Analogic

(508) 977-3000

Fax: (508) 977-6813

www.analogic.com

Speech recognition and text-to-speech systems. Also new is their IP-Voice, a technology that sends voice over IP networks, bypassing the public network and its annoying tolls.

Angus Telemanagement Group

(905) 686-5050 (800) 263-4415

Fax: (905) 686-2655

www.angustel.ca

AnswerSoft

(972) 997-8300

www.answersoft.com

Sixth Sense computer telephony automation software. They also offer a variety of call center applications built on their CT platform: for example, a routing app that automatically retrieves ad routing customer data with calls. New: Concerto, an Internet/intranet CTI system, along with several other flavors of Inter/intranet product.

AnswerThink

(770) 690-9700

Fax: (770) 690-9710

www.answerthink.com/

APAC TeleServices

(800) 688-7987

Fax: (847) 945-2938

www.apacteleservices.com

One of the country's largest service bureaus. They also own an 800 number that must make their competitors gnash their teeth in fury: 800-OUTSOURCE. Services offered include: technical support and pre-sale services; marketing response tracking and media sourcing; order capture and processing; up-selling and cross-selling; and financial transaction services.

Apex Voice Communications

(818) 379-8400 (800) 727-3970

Fax: (818) 379-8410

www.apexvoice.com

Specializes in high-density, scalable systems for Enhanced Services, call centers and CTI apps. systems for network service operators/telephone companies, OEMs and VARs. Products include: OmniVox, an app gen; OmniNet, a Web-accessible network services manager; APEX Messaging System, a messaging system for telephone network operators (wireless/wireline).

Applied Innovation Management

(510) 226-2727 (800) 942-7754

Fax: (510) 226-7990

www.aim-helpdesk.com

Applied Innovation Management's help desk product Online SupportCenter is a web-based software system for managing external product support. It's a combination of what you'd expect from classic help desk (a problem resolution engine, customer and call tracking, etc.) with a web front end.

Applied Voice Technology

(425) 820-6000

Fax: (425) 820-4040

www.avtc.com

Makes a line of "open systems based advanced computer telephony products," specializing in fax and call processing, unified messaging, and call centerproductivity and customer service.

Applix

(508) 870-0300

www.applix.com

Appro International

(408) 452-9200 (800) 927-5464

Fax: (408) 452-9210

www.appro.com

Industrial grade computer platforms for critical applications. Like call centers. Includes hot swap fans, hot swap power supplies, temperature monitoring systems — all the good stuff.

Apropos Technology

(630) 472-9600

Fax: (630) 472-9745

www.apropos.com

Apropos' latest version of their Call Link Server system (version 3.5), this time ported to NT and "re-architected" to take advantage of some of the features of that platform. For example, the administrative components of Call Link are implemented using the Microsoft Internet Information Server built into NT Server 4.0. That allows Call Link to be administered from any Java-supporting browser. New in version 3.5 is messaging middleware that supports over 1,000 connections, which reportedly supports call centers of as many as 500 to 600 agents within a single Call Link system. Scripts created for the previous OS/2 version of Call Link are "nearly 100%" compatible with the new system (seems they are leaving a little wiggle room there), and they promise a fast conversion. It costs $40,000 for a five agent system with one supervisor, a report system and a server.

Aptex Software

(619) 623-0554 (888) 623-0554

Fax: (619) 623-0558

www.aptex.com

Ara Training

+1 (47) 23039600

Fax: +1 (47) 23039601

www.ara.no

Delivers training for call centers in Europe in english and the scandinavian languages.

ArelNet

(212) 935-3110

Fax: (212) 935-3882

www.arelnet.com

Arial Systems

(847) 573-9925

Fax: (847) 573-9926

www.arialsystems.com

ArialView system: ceiling-mounted "nodes" receive ID signals from infrared transmitters embedded within personnel badges and tags. As people move around a facility, the system knows where they are. Calls can be sent to the nearest phone set.

Ariel Corporation

(609) 860-2900

Fax: (609) 860-1155

www.ariel.com

Designs, manufactures and markets DSP-based data communications hardware and software products. The company's Rascal and T-1 Modem product families are remote access and modem pool solutions providing ISDN and 56-kbps modem remote access.

Aristacom

(510) 748-1564
Fax: (510) 748-1534
www.aristacom.com
CTI middleware software.

Artisoft

(617) 354-0600 (800) 914-9985
Fax: (617) 354-7744
www.artisoft.com
Telephony systems including TeleVantage, a PC-based phone system with call center features including "group call distribution", Visual Voice, a telephony toolkit for creating Windows-based applications, and InfoFast, a a data retrieval tool for fulfilling customer info requests.

Aspect Telecommunications

(408) 325-2200 (800) 541-7799
Fax: (408) 451-2746
www.aspect.com
CallCenter ACD, Agility voice processing system, EnterpriseAccess software for linkage between applications. Aspect's switches are recognized as high-quality, high-volume ACDs.

Association for Services Management Internatio

(941) 275-7887 (800) 333-9786
Fax: (941) 275-0794
www.afsmi.org
An industry association of executives, managers and professionals in the services and support industry.

Astea

(617) 275-5440
Fax: (617) 275-1910
www.astea.com
Customer Enterprise Series (PowerHelp, PowerSales, Dispatch-1) for help desks. Strong in the field service and internal help desk environment. They also offer the Heat help desk system, acquired through Astea's acquisition of Bendata.

AT&T

(800) 222-0400

Fax: (908) 221-3188

www.att.com/business/global

Long distance, toll-free and call center consulting. They also feature a variety of network-intelligence call routing products and service offerings. What else can you say about one of the largest corporations in the world? They have a huge impact on the call center industry, and have an enormous well of expertise on the subject.

Atio Corporation

(612) 837-4055

www.atio.com

CyberCall is a modular call center solution. The modules include Inboundtelephone call, IVR (interactive voice response), e-mail, fax and webcapabilities (web callback, text chat and Voice over IP).

AuBeta Telecom

(425) 869-1700

www.aubeta.com

AuBeta Telecom's iQueCall is an out-of-the-box family of customer focused call center systemsfor NT and IP networks. IQueCall integrates with host of client/server databases for single- or multi-site call centers with between 5 and 300 agents. Functions supported include; CTI/PBX interface, SQL database interface, ACD, Universal Messaging ACD, host interface, IVR, predictive dialer, faxing, letter management, documentation access, problem resolution, agent monitoring, call recording, on-screen prompting and followup.

Aurora Systems

(508) 263-4141

Fax: (508) 635-9756

www.fastcall.com

FastCall computer telephony software (middleware). Allows you to hook client/server and LAN apps to your phone system through TAPI and/or TSAPI standards. The product is sold exclusively through distribution partners (companies like Nortel, Norstan, Harris, etc.).

Aurum Software

(408) 986-8100 (800) 683-8855

Fax: (408) 654-3400

www.aurum.com

Aurum's Customer Enterprise is an integrated suite of applications that encompasses a company's enterprise-wide processes of field sales, channel sales, telesales, telemarketing, corporate marketing and customer service. The Aurum Customer Enterprise implements a "closed-loop" flow of information, automating and integrating "front office" sales, marketing, and customer service functions with "back office" finance, manufacturing, and order entry applications. The suite is modular, built around a set of components for each application.

Automatic Answer, The
(714) 661-2660 (800) 410-2745
Fax: (714) 661-0778
www.taa.com
Sells automated attendants and voice processing systems based on industry-standard PC platforms.

Balisoft
(416) 256-1419
Fax: (416) 256-0706
www.balisoft.com
LiveContact, a web/call center integration suite that has collaborative tools, as well as text-chat and real-time agent connections.

Banksoft
(714) 975-0796
www.banksoft.net
Small call center system called VoiceSolution, which (under Windows platforms) provides call processing and backend data processing with the major industry-standard databases. They'll do software-only, hardware-only, or a turnkey package.

Bard Technologies
(800) 997-4470
Fax: (914) 234-3028
www.bardtech.com
CallLab ACD Simulator. System that models activity in a call center, projecting future activity (in aggregate) and allowing you to experiment with changing call center parameters. A good way to model the behavior of skills-based routing (and other "random" activities that bedevil Erlang calculations).

Barnhill Associates
(303) 792-9500
Fax: (303) 792-5800
www.barnhill.com
Systems integrators and consultants on "process reengineering."

BCS Technologies
(303) 713-3000
Fax: (303) 713-3030

www.bcstechnologies.com

PBX/ACD for small call centers. The DSP1000 is an open architecture, standards-compliant switch with standard routing capabilities and both real-time and historical reporting. BCS also provides systems for CTI, skills-based routing, intelligent announcements and "3-D graphic displays and reports" for call centers from 5 to 750 agents.

Belgacom

(203) 221-5280

Fax: (203) 222-8401

The Belgian national telecom carrier, which is actively engaged in assisting companies in locating their call center operations in that country.

Bendata, Inc.

(719) 531-5007 (800) 776-7889

Fax: (719) 536-0620

www.bendata.com

Help desk software; HEAT (HelpDesk Expert Automation Tool) for Windows is a system for call tracking, problem resolution and messaging, management reporting and trend tracking.

Berkeley Speech Technologies

(510) 841-5083

Fax: (510) 841-5093

www.bestspeech.com

BeSTspeech text-to-speech conversion technology. Works with all major app gens, board lines, operating systems.

Bicom

(800) 766-3573

Fax: (203) 268-3404

Blue Pumpkin Software

(650) 948-4998

Fax: (650) 948-4082

www.blue-pumpkin.com

PrimeTime workforce management software. PrimeTime's advantage in this market is that it incorporates the most important features needed to get a workforce schedule up and running, at a much lower price point that the others (about $25,000 for a small center, roughly $1,000 a seat). The Win 95 or NT product is the easiest scheduler to use that I've seen. A supervisor can put a roster together in minutes, play what-if scenarios, and use templates to make it fast on an ongoing basis. In order to make it fast and cheap, BP sacrificed some of the features others offer: adherence, for one, and historical projection. It won't go looking at three years of records to predict what your volume will be next Friday. But if you know what your volume is likely to be in various scenarios, you can set up templates to identify each kind of week that occurs. Blue Pumpkin's software will be sold by Teloquent as an optional enhancement to their call routing product. I think this product will shake up the moribund product category — it focuses on what call center managers actually want, acts as an aid to what they already have to do, without forcing them to swallow (and pay through the nose for) a lot of useless features. This is scheduling made easy.

Bogen Communications
(201) 934-8500
Fax: (201) 934-9832
Messages-on-hold, digital announcers, voice loggers and recorders.

Bosch Telecom
(908) 769-8700
Fax: (908) 226-8787

Boston Communications Group
(617) 692-7000
www.bgci.net
Call center outsourcing services, particularly to the wireless telecom and prepaid telecom industries.

Boyd Company
(609) 890-0726
Consultant specializing in call center location and site selection.

Bramic Creative Business Products
(905) 649-2732
Fax: (905) 649-2734
Ergonomic furniture for call centers. (People tend to forget how important this stuff is.)

Brightware
(415) 884-4744
Fax: (415) 884-4740
www.brightware.com
Web-based customer interaction systems; software that automatically answers customers' web and e-mail inquiries and routes those that need agent attention to the right rep.

Bristol Group, The
(415) 925-9250

Fax: (415) 925-9278

www.bg.com

Large scale faxing systems, with software that runs on a Sun platform.

Brite Voice Systems

(407) 357-1000

www.brite.com

Voice processing and IVR systems that integrate voice, fax, CTI and Internet capabilities.

Broadbase Information Systems

(800) 966-8085

Fax: (650) 614-8301

www.broadbase.com

Enterprise Performance Management (EPM) systems — analytic applications that optimize business performance in the enterprise.

Brooktrout Technology

(781) 449-4100

Fax: (781) 449-3171

www.brooktrout.com

Fax, voice and telephony products, mainly at the component level. Known best for the TR series of fax and voice boards, and for Show N Tel, a voice processing/IVR platform.

Buchanan E-Mail Limited

+44 (19756) 51741

Fax: +44 (171) 681 1218

www.buchanan.co.uk

Buffalo International

(914) 747-8500

Fax: (914) 747-8595

www.buffalo-intl.com

OAPDE (Open Architecture Predictive Dialing Environment), PC-based outbound dialer. A strong platform for buildng a LAN-based dialing and call processing system, at a relatively low per-seat cost (relative, of course, to a full-featured turnkey dialer).

BWT Associates

(508) 845-6000

Fax: (508) 842-2585

Disaster recovery consulting and services.

CACI Products Company

(619) 824-5200

Fax: (619) 457-1184

www.caciasl.com

Makes the Call Center Maximizer, a tool that lets you set up a model of staffing and volume, add routing and other parameters, and play out various scenarios to make it most efficient.

Call Center Network Group

(817) 275-5853 (800) 840-2264

Fax: (817) 275-5530

www.ccng.com

Membership organization for call center professionals.

Call Center Solutions

Fax: (203) 740-3536

www.callcenters.com

Predictive dialers and call blending equipment.

Call Center University

(615) 221-6850

Fax: (615) 221-6885

www.callcenteru.com

Professional organization created as a separate business unit of TCS Management Group to promote certification programs and call center management standards. The CCU Certification Program provides certification to participants who complete a series of courses and assessments to help them acquire new knowledge and skills for improving their centers. And they offer Symposium series of two-day seminars for updating call center professionals on current issues and technologies.

Call Interactive

(800) 428-2400

www.callit.com

IVR and call center outsourcing.

Call One

(800) 749-3160

Fax: (407) 799-9222

www.call-1.com

Distributes headsets, conferencing equipment.

Call Ontario - Bell Canada

(800) 917-1917

Fax: (416) 971-6581

CallCenter Technology

(203) 881-2869

Fax: (203) 840-0209

www.callcentertechnology.com

Makes a supervisor and call center knowledge management tool that lets users combine and view at the desktop, multiple data sources within the call center. ACD statistics, workforce management data, quality performance scores and other call center data can be monitored, modified and reported in any format the user desires. Thresholds can be set on all data to alert users of potential problems and event status is uniquely displayed with the call center's floor plans making it easy to spot location requiring assistance.

Callscan Australia

Fax: +61 (3) 9253 1099

www.callscan.com.au

Provider of call center products and services to the Australian and NewZealand markets.

CallWare

(801) 486-9922 (800) 888-4226

Fax: (801) 486-8294

www.callware.com

Software for integrating voice, fax, e-mail, IVR.

Calonge & Associates

(210) 497-2320

Fax: (210) 497-4116

www.ca-script.com

Scripting and script consulting for business-to-business and business-to-consumer marketing campaigns.

Cardinal Technologies

(800) 775-0899

Fax: (717) 293-3055

Carnegie Group

(412) 642-6900

www.cgi.com

Consulting, application development and systems integration for call centers. Also recently bought Advantage kbs. Then Carnegie was itself bought by Logica Inc.

CAS Marketing

(402) 393-0313

(800) 524-0908

Fax: (402) 390-9497

List management services, as well as phone number appending and other lookup services.

Cascade Technologies

(732) 906-2020

Fax: (732) 906-2018

www.cascadetechnologies.com

Cascade makes software mainly for employee benefits and human resources, like record-keeping systems for defined contribution plan administration. The company also develops and markets a suite of interactive communication systems designed to automate workflow and employee service requests using voice response, fax, email, the Internet and employee call centers.

Castelle

(408) 496-0474 (800) 289-7555

Fax: (408) 492-1964

www.castelle.com Fax systems.

CCMA Asia Pacific

+61 (4) 18 257 252

CellIT

(305) 436-2300

UnPBX

CenterCore

(908) 561-7662 (800) 220-5235

Fax: (908) 561-0911

www.centercore.com

Cubicles and agent workstations (the physical desks, not the computers on the desks) for call centers. They make a space-saving cluster design. And their modular workstations are fully equipped to hold (and hide) wiring, etc.

CenterForce Technologies

(301) 718-2955

Fax: (301) 718-3667

www.cforcetech.com/

Call center optimization system for outbound campaigns.

Centigram

(408) 944-0250

Fax: (408) 428-3722

www.centigram.com

Voice processing equipment. Makes the Series 6 communications server, which is the basic IVR platform, and a multimedia messaging product called OneView.

Century Telecommunications

(512) 353-1155 (888) 888-8757

Fax: (512) 754-5678

www.cticallcenter.com

Call center outsourcer that began life as the excess operator services capacity of a Texas telco.

Chadbourn Marcath

(312) 915-0300 800-ACD-STAT

Fax: (312) 915-0366

CCAnalyzer, call center management software. Software for displaying center stats to agents that runs on tv monitors, not readerboards.

ChiCor

(312) 322-0150 800-448-TRPS

Software for disaster planning and recovery.

Chordiant Software

(650) 493-3800 (888) 246-7342

Fax: (650) 493-2215

Software for managing customer data, including transaction data, customer histories, and the business processes that determine how you are going to handle any given transaction.

Cicat Networks

(703) 383-1400

Fax: (703) 385-3788

www.cicat.com

ISDN specialists, including line ordering and help with provisioning for ISDN equipment.

Cintech

(513) 731-6000 (800) 833-3900

Fax: (513) 731-6200

www.cintech-cti.com

Cintech is a provider of ACD for small and mid-size call centers. Departments, branch offices, small offices and informal environments can reap the benefits of ACD with their family of solutions for users of Nortel's Norstar and NEC's NEAX 2000 IVS switches.

Clarify

(408) 428-2000

Fax: (408) 428-0633

www.clarify.com

Help desk software that has moved beyond backend problem processing into that fuzzy realm where they can claim to being a total, enterprise-wide sales and service tool.

ClearVox Communications

(408) 371-8400

Fax: (408) 559-6760

www.clearvox.com

ClearVox manufactures hands-free headsets for PCs, cordless phones, the Internet and other applications.

Clientele Software

(503) 612-2773

Fax: (503) 612-2973

www.clientele.com

Customer service software for organizing, tracking and sharing customer information from the agent workstation.

Com2001 Technologies

(760) 431-3133

Fax: (760) 431-3141

www.com2001.com

CoMatrix

(714) 992-5982 (800) 888-7822

Fax: (714) 992-5980

Supplier of used telecom equipment from major vendors, including Norstar, AT&T and Toshiba. They also distribute for Voysys and other voice mail systems.

Comdial

(804) 978-2200 (800) 347-1432

Fax: (804) 978-2230

www.comdial.com

LAN-based ACD through their QuickQ software product, plus a variety of interesting software and combo products that bring advanced features closer to the middle end of the market.

Comdisco Disaster Recovery

(800) 272-9792

Fax: (847) 518-5340

www.comdisco.com

Disaster recovery and service assurance programs. They can literally set up an entire center for you if something bad happens (earthquake, flood, etc.). Or something less extensive if your problem is less dramatic.

CommercePath

(800) 600-4329

www.commercepath.com

An EDI-to-fax system that extends a company's EDI to suppliers and trading partners not equipped for that transaction system. Quite useful, if a bit specialized. CommercePath is a subsidiary of Applied Voice Technology.

Commetrex

(770) 449-7775

Fax: (770) 242-7353

www.commetrex.com

Computer telephony boards, particularly the MSP ("Media Stream Processor") DSP resource board. These boards support voice, fax, speech recognition, etc. for call center applications.

Communication Service Centers

(800) 537-8000

Communications Managers Association

(973) 425-1700 (800) 867-8008

Fax: (973) 425-0777

www.cma.org/

CommuniTech

(847) 439-4333

Fax: (800) 783-7800

www.communitech.com

One of the largest distributors of headsets around. Stocks ACS, UNEX and VXI. Also is the sister company of CommuniTech Services, a systems integrator.

Comport Distributions

(800) 789-0090

Fax: (516) 473-2246

Computer Communications Specialists

(770) 441-3114 (800) 227-7227

Fax: (770) 263-0487

www.ccsivr.com

A hardware and software combo platform for IVR called FirstLine.

Computer Sciences Corporation

(617) 332-3900

Fax: (617) 332-2864

www.csc.com

Computer Talk Technology

(800) 410-1051

Fax: (905) 882-5000

www.icescape.com

Server-based ACD (with digital switching and built-in CTI), and a newly announced Internet ACD.

Comverse Information Systems

(516) 677-7400 (800) 967-1028

Fax: (516) 677-7399

www.cis.comverse.com

Digital recorders and voice loggers, and monitoring systems.

Connected Systems

(805) 962-5066

Fax: (805) 962-5044

www.connectedsystems.com

Conner Group

(770) 622-3987

Fax: (770) 622-3987

Consolidated Market Response

(800) 500-6006

Fax: (217) 348-7060

Inbound and outbound telemarketing and teleservices, including backend stuff like fulfillment and database management.

Contact Dynamics

(312) 345-1344

Fax: (312) 345-1880

www.contactdynamics.com

Software and services for interactive internet communications; tracking of visitors to a web site, text-based communication between the rep and the customer.

Conversational Voice Technologies/ConServIT
(847) 249-5560 888-343-CVTC
Fax: (847) 249-9773
www.cvtc.com
Inbound service bureau.

Copia International
(630) 682-8898 (800) 689-8898
Fax: (630) 665-9841
www.copia.com
They make the FastFax fax server engine, a high volume network-based fax system. Add-ons and options include broadcast faxing, internet faxing and international products.

CoreSoft Technologies
(801) 431-0070 (877) 611-9090 x3
Fax: (801) 431-0079
www.coresoft.com
Multi-function telephony equipment that purports to connect all the disparate communication devices (fax, phone, pagers, etc.) into one environment.

Cortelco
(901) 365-7774 (800) 866-8880
Fax: (901) 365-3762
www.cortelco.com
Switching systems, ISDN equipment, and software to manage it.

CosmoCom
(516) 851-0100
Fax: (516) 851-1005
www.cosmocom.com
CosmoCom offers live, one-to-one, integrated multimedia customer service for Internet and telephone callers by combining the functions of ACD, CTI, IVR and unified messaging in one totally IP-based system.

Crystal Group, Inc.
(319) 378-1636 (800) 378-1636
Fax: (319) 393-2338
www.crystalpc.com
Industrial grade "fault resilient computer systems, the kind that keep mission critical applications like call centers running all the time.

CSI-Data Collection Resources

(860) 289-2151

Fax: (860) 289-4662

www.csiworld.com/dcr

CSI-Data Collection Resources is a developer of automated voice and data products for call centers. Products include: "ACD Data Collection and Reporting System" and "ACD Agent Monitoring System."

CT Solutions

(732) 360-2863 (800) 724-0802

Fax: (732) 360-2879

www.solutions4ct.com

Computer telephony equipment VAR for a variety of manufacturers.

CTL

(203) 925-4266

Fax: (203) 925-4267

www.ctlinc.com

VoiceSupport voice processing system for the low end of the market (2 ports) and the Interactive Support IVR system for higher end users.

Curasoft

Fax: (510) 795-6109

www.curasoft.com

Customer Support Consortium

(206) 622-5200

Fax: (206) 667-9181

www.customersupport.org

Industry association to develop techniques and spread information for the service and support call center professional.

Dakotah Direct

(509) 789-4500 (800) 433-3633

Fax: (509) 789-4706

www.dakotahdirect.com

Outsourcing call center service bureau specializing in outbound, inbound and "comprehensive account management". Industries served include telecom, finance, insurance, high tech and utilities.Dakotah Direct designs and implements nationwide teleservices for national clients in the financial, telecommunications, insurance, utilities, and high-tech industries.

Daktronics

(605) 697-4468 (888) 325-8766

Fax: (605) 697-4700

www.daktronics.com

InfoNet multiline readerboards with custom and standard interfaces.

Data Processing Resources Corporation

(909) 444-9476

Fax: (909) 444-9269

www.dprc.com

Their Call Center and Telecommunications Practice focuses on the applications and technologies used in call centers, network, and telecommunication organizations. DPRC develops strategic and technology plans; designs, develops and implements applications, networks, and systems; and provides management and operations support for call center systems, networking solutions, and telecommunication systems.

Database America Companies

(201) 476-2300

www.databaseamerica.com

One of the largest database and list management firms. They recently merged with American Business Information to form one of the most fully-featured data companies in the industry.

Database Systems

(602) 265-5968 (800) 480-3282 Fax: (602) 264-6724 www.dsc1.com TeleMation telemarketing software.

Datamonitor

(212) 686-7400

Fax: (212) 686-2626

www.datamonitor.com

Datapoint

(210) 593-7000 (800) 733-1500

Fax: (210) 593-7518

www.datapoint.com

Computer-based communications solutions, including client-server, video communications and integrated telephony applications.

Davox

(978) 952-0200

Fax: (978) 952-0201

www.davox.com

Predictive dialers, as well as a CTI/blend system called SCALE and a dedicated inbound product called In/Unison. Also offers the Unison suite of call center management solutions incorporating predictive dialer systems, inbound CTI solutions, and advanced integrated inbound/outbound call blending systems compatible with major ACDs and PBXs. Davox systems are used worldwide in collections, telemarketing, telesales, customer service, fund raising and other mission-critical customer contact environments.

Decisif Software Solutions
(514) 362-7117 (888) 517-2929

Fax: (514) 362-0456

www.decisif.com

Decisif provides a suite of CTI applications such as Intelligent CallRouting, Help Desk Automation, Call Tracing and Reporting, PredictiveDialing, Internet Integration and Interactive Voice Response.

DecisionKey
(603) 622-8080 (888) 609-5980

Fax: (603) 622-8282

www.decisionkey.com

Expert system problem resolution engine for support call centers.

Dees Communications
(800) 663-5601

Fax: (604) 946-8315

Universe Centrex ACD for Nortel switches. Mediator Direct ACD, routing system for centrex.

DialAmerica Marketing
(201) 327-0200 (800) 526-4679

www.dialamerica.com

Telemarketing outsourcing firm.

Dialogic
(973) 993-3000 (800) 755-4444

Fax: (973) 993-3093

www.dialogic.com

Voice cards, SCSA hardware, GammaLink fax boards. CT-Connect CTI Software accessible through HP's Smart ContAct.

Dictaphone
(800) 447-7749

Fax: (203) 386-8595

www.dictaphone.com

Dictaphone's Symphony CTI is an advanced communications management system used by call centers to record and archive the tons of calls they take or make. It integrates with a call center's switch to capture real-time data for enhanced searching and customized recording solutions.

Digisoft Computers

(212) 687-1810

Fax: (212) 687-1781

www.digisoft.com

Telescript, call center software for PC based networks, manages multiple inbound/outbound/blended projects simultaneously. Features: scripting with advanced logic & branching; inbound with ANI/DNIS; predictive dialing.

Digital Software International, Inc

(760) 741-8800

Fax: (760) 741-2693

www.digisoftware.com

Centergy Scriptbook allows non-programmers to create scripts for useby agents in outbound or inbound call centers. Unlimited branching, seamless interfacing to foreign databases and programs through ODBC and OLE. The script on the agent's desk appears like a "notebook" with tabs. As theagent enters data on the script, it automatically branches to the nextcorrect scriptpage. Digital Software also has software products for monitoring agent productivity and supervision.

Direct Marketing Association

(212) 768-7277

Fax: (212) 768-4546

www.the-dma.org

Industry association focusing on call centers as an offshoot of direct marketing. When they get involved it is mostly from the point of view of outbound telemarketing. They run many seminars and shows all over the US.

Direct Marketing Solutions

(800) 688-2367

Discovery Training & Development

(602) 258-9500

Fax: (602) 258-9555

www.discoverytraining.com

Distributed Bits

(312) 207-1500

Fax: (312) 207-0041

www.dbits.com

Makes an e-mail tracker and response automator for call center use called ResponseNow.

DP Solutions

(610) 317-9181

Fax: (610) 317-9491

www.dpsol.com

Help Desk and customer support software.

DPC Computers

(914) 426-3790

Fax: (914) 426-6275

www.salestax.com

They make a useful and inexpensive piece of software that shows sales taxes for any locality during a telephone transaction. A must for the small center.

Drextec

(609) 596-8285

Fax: (609) 596-8447

www.drextec.com

A PC-LAN-based open predictive dialing and telemarketing software system, for call centers of 12 to 144 agents. Based on the Buffalo International OADPC dialing engine.

DSP Group

(408) 986-4300

Fax: (408) 986-4490

www.dspg.com

Speech compression technology, optimized particularly for conferencing and internet telephony applications.

Dytel

(847) 981-1200

Fax: (847) 439-3456

www.dytel.com

Automated attendants.

E-Voice Communications

(408) 991-9988

Fax: (408) 991-9630

www.evoicecomm.com

E-Voice Communications develops voicemail systems that incorporate unified messaging technology. Their systems are based on the open Windows 95/NT platforms.

Easyphone SA

011-351-1720-5050

Fax: 011-351-1720-5090

www.easyphone.pt

Portuguese firm that makes the CallPath-based EasyPhone Call Center Management Software system for NT. Incorporates CTI, software-based predictive dialing and scripting/database features. Worth a look.

EasyRun

(201) 541-1855

Fax: (201) 541-8333

www.easyrun.com

Developer of desktop computer telephony systems. Products include: ECC, a suite of call center applications featuring integrated IVR, voice mail and skills-based routing. It provides MIS application for real-time and historical reports, agent station application and readerboards. ETSC is a telephony switching system running on a network of desktop PCs. It provides the entire feature set required by a mobile switching system (MSC) and follows the GSM standards. The system works with a 'dumb-PBX' to provide a complete switching platform for specialized telephony.

Edgewater Technologies

(781) 246-3343

Fax: (800) 233-7997

UnPBX's. (A k a, the WhiteCap communications server system.)

Edify

(408) 982-2000 (800) 944-0056

Fax: (408) 982-0777

www.edify.com

IVR and workflow software for backoffice processing of call center transactions.

Edward Blank Associates

(212) 741-8133

Inbound and outbound service bureau.

eFusion

(503) 207-6300

Fax: (503) 207-6500

www.efusion.com
An "Interactive Web System" for self-service that allows customers to retrieve information in a way that mimics IVR interaction. It's a platform for application creation (they cite "Internet Call Waiting" as one possibility) that exploit combined voice and data networks for enhanced customer interactions and e-commerce.

eGain Communications Corp.
(408) 737-8400 (800) 603-4246
www.egain.com
EGainTM develops customer service solutions for electronic commerce.

EIS International
(703) 478-9808 (800) 274-5676
Fax: (703) 326-6621
www.eisi.com Predictive dialers.
Outbound and integrated inbound/outbound applications for call centers. EIS provides systems for telemarketing, customer service, fund-raising, market research and collections. EIS recently adopted CORBA (Common Object Request Broker Architecture) as a platform for future products.

Energy Enterprises
877-74ENERGY
Fax: (714) 525-6716
energyenterprises.com
Call center training and consulting services.

Engel Picasso Associates
(800) 241-8092

Enterprise Computer Telephony Forum (ECTF)
(510) 608-5915
Fax: (510) 608-5917
www.ectf.org
Vendor-membership organization for developing interoperability standards among computer telephony devices.

Enterprise Integration Group
(510) 328-1300
Fax: (510) 328-1313
www.eiginc.com
Services geared to CTI; things like project management, IVR design and testing, business process analysis, training, etc.

Entertainment Technology

(416) 598-2223

Fax: (416) 598-5374

Call center display board system (called FRED) that actually works with TV monitors; in addition to displaying ACD stats and other traditional kinds of data, it also allows you to mix in video and other kinds of messages.

Envision Telephony

(206) 621-9384

Fax: (206) 621-7525

www.envisiontelephony.com

Makes a product called SoundByte Enterprise 97, which is monitoring and quality assurance software for call centers. Win 95/NT platform, it's composed of three parts: SoundByte Server, Supervisor and Agent. Used by major centers like those in Hewlett-Packard, Nordstrom and Harte Hanks.

Envox

(941) 353-6587 (888) 368-6987

www.envox.com

Envox Script Editor for developing multimedia call center and CT apps.

Epigraphx

(650) 802-5858

Fax: (650) 802-5850

www.epigraphx.com

Fax-on-demand and, lately, internet-enabled fax/web combos that call centers can use to further reduce call volume.

eShare Technologies

(516) 864-4700 (888) 374-2734

www.eshare.com

Web-based customer service and support, customer self-service, live conferencing and events, distance learning, chat, threaded discussion forums, and custom integration tools.

Estech

(972) 422-9700

Fax: (972) 422-9705

Makes the IVX telephone/voice mail product. Not really a call center product except at the very low end, if you need something that will make a departmental center with rudimentary ACD features.

Exacom

(603) 228-0706

Fax: (603) 228-0254

www.exacomusa.com

Automated messaging system (MessageMaxx); digital recording systems.

Exchange Applications

(617) 737-2244

Fax: (617) 443-9143

www.exapps.com/

Executive Call Centers

(800) 888-3188

Executone Information Systems

(203) 876-7600 (800) 808-1305

Fax: (203) 882-0400

www.executone.com

Integrated Digital System platform. ACDs, predictive dialers.

Expert Systems

(770) 642-7575

Fax: (770) 587-5547

www.easey.com

Ease IVR development product. Latest platform supported is Windows NT.

Exsys

(505) 256-8356

Fax: (505) 256-8359

Expert system software and services.

Far Systems

(920) 563-2221 (800) 805-0082

Fax: (920) 563-1865

www.farsystems.com

Interactive voice response application generation software and systems.

Fax2Net

(301) 721-0400

Fax: (301) 721-0083

www.fax2net.com

FaxNet

(617) 557-4300

Fax: (617) 557-4301

www.faxnet.com

Enhanced fax services, including long distance service, that specializes in delivery of fax over IP networks (including the Internet).

FaxSav

(732) 906-2000 (800) 828-7115 x2330

Fax: (732) 906-1008

www.faxsav.com

Internet fax systems.

FaxStar

(714) 724-0806 (800) 327-9859

Fax: (714) 752-7980

FaxStar, enterprise-wide fax server system.

Figment Technologies

(416) 225-7913 (888) 292-5089

Fax: (416) 225-7845

www.unimessage.com

UniMessage, a unified messaging product.

France Telecom

www.francetelecomna.com

French national carrier, wants to attract call centers to that country.

Franklin Telecom

(805) 370-0409 (800) 372-6556

Fax: (805) 373-7373

www.ftel.com

Systems for voice-over-internet communications.

Frost & Sullivan

(650) 961-9000

Fax: (650) 961-5042

www.frost.com

Market research on technology, much of it focused on IT, telecom and call centers.

Fujitsu Business Communication Systems

888-FBCS-CTI

Fax: (602) 921-4800

www.fbcs.fujitsu.com

Offers a wide variety of call center tools, starting with core switches and moving up through CTI links and specific applications and services.

Funk Software

(617) 497-6339 (800) 828-4146

Fax: (617) 547-1031

www.funk.com

Remote control technology that incorporates remote control, screen monitoring, and screen record and playback, for help desk, network administration, call center, and training applications.

Fuseworks

(819) 771-8182

Fax: (819) 771-5258

www.fuseworks.com

Fuseworks offers a "live" Internet marketing solution: Call center agentscan track Web site visitors in real time, match profiles againsthistorical data, and engage them in collaborative multimediasessions.

Gannett Telematch

(800) 523-7346

Fax: (703) 658-8301

www.telematch.com

Telematch provides database marketing services which include residential and business telephone number appending, data enhancement, database management, data processing, and laser printing services.

GBH Distributing

(818) 246-9900 (800) 222-5424

Fax: (818) 246-5850

www.gbhinc.com

Headsets for a variety of applications, including call centers.

Genesys Telecommunications

(415) 437-1156 888-GENESYS

Fax: (415) 437-1260

www.genesyslab.com

Call Center Manager: Combined inbound/outbound call processor. And the Genesys T-Server, a CTI server that provides connectivity between data and telephone networks. Through Genesys applications, you can put call center features on your network, like skills-based routing, or predictive dialing.

GeoTel

(978) 275-5149

Fax: (978) 275-5399

www.geotel.com

Intelligent CallRouter (ICR), network-based ACD router. They have one of the most interesting ways of implementing virtual call centers (but by no means the only one).

GM Productions

800-827-DEMO

Fax: (770) 237-5522

www.gmpvoices.com

Professional recording of voice prompts and other kinds of announcements, including on-hold systems.

GN Netcom/Unex

(603) 598-0488 (800) 345-8639

Fax: (603) 598-1122

www.gnnetcom.com

Headsets.

Graybar

(314) 512-9200 (800) 825-5517

www.graybar.com

Distributor of telecom and call center products along a wide spectrum of technologies, from a lot of manufacturers. Their catalog is worth looking at.

Greater Houston Partnership

(713) 844-3600

Fax: (713) 844-0200

www.houston.org

Organization of government and business resources for companies that are locating in Houston, with an eye to business development and attracting call centers to that area.

Group 1 Software

(800) 368-5806

Fax: (301) 731-0360

www.g1.com

Software for mailing and other direct marketing efforts.

Harris

(415) 382-5000 (800) 888-3763

Fax: (415) 883-1626

www.harris.com

Switching systems & PBXs, particularly VoiceFrame, a programmable switch with an NT server.

Harrison Direct, Inc.

(423) 867-8209

Fax: (423) 867-8523

www.harrisondirect.com

Harrison Direct, Inc. is a direct marketing company working Fortune 1000 companies. HDI's services are structured "to enhance the brand of the strategic business partner in the mind of the consumer."

Hello Direct

(408) 972-8155 (800) 444-3556

www.hello-direct.com

Major catalog distributor of headsets and other consumer telephony devices.

Help Desk Institute

(800) 248-5667

Fax: (719) 528-4250

www.helpdeskinst.com

Industry organization devoted to the inbound, mainly support-based, call center. They do a fairly good amount of targeted research, publish some interesting studies, and are generally pretty helpful.

Hewlett-Packard

(650) 857-1501

Fax: (650) 857-5518

www.hp.com/

HTL Telemanagement

(301) 236-0780

Fax: (301) 421-9513

www.htlt.com

Hills B calculator for simulating call center conditions.

Human Resources Dimensions

(817) 801-5100

Fax: (817) 801-9079

www.hrdimensions.com

Specializes in filling management positions for call centers across all functional areas for clients engaged in inbound or outbound customer service, telemarketing, credit & collections, telesales, help desk, fulfillment, verification, etc.

I-BUS
(619) 974-8426 (800) 382-4229
www.ibus.com
OEM PC manufacturer.

IBM/Early Cloud
(401) 849-0500 (800) 322-3042
Fax: (401) 849-1190
www.earlycloud.com
CallFlow, distributed software for large-scale call center automation that allows companies to automate customer contact applications such as customer service, telesales, account management and collections. CallFlow lets you build scalable applications in high transaction volume client/server environments.CallFlow provides application generation, business workflow, computer telephony integration, contact management, fulfillment and call result reporting. It enables access to corporate data systems throughout the enterprise through a unique object-driven technology.

ICM Conferences
(312) 540-5693
Fax: (312) 540-5690

ICT Group
(215) 757-0200
Fax: (215) 757-4538
www.ictgroup.com
Telemarketing service bureau.

IEX
(972) 301-1300
Fax: (972) 301-1200
www.iex.com
Call center management software, workforce management software.

Imagine (USA) Ltd.
Fax: (770) 921-6271
www.imagineusa.com

Impact Telemarketing Group
(609) 384-1111 (800) 793-2345
Fax: (609) 853-6859
www.impacttele.com
Inbound and outbound teleservices, including account management, outbound telemarketing, accounts receivables, mailing and fulfillment services.

Incoming Calls Management Institute

(410) 267-0700 (800) 672-6177

Fax: (410) 267-0962

www.incoming.com

Training seminars, consulting, newsletter and other publications for call center managers.

Industrial Computer Source

Fax: (619) 677-0615

Manufactures rugged PC- compatible computers, rack-mount chassis, I/Oand networking for computer telephony, factory automation and processcontrol, scientific and engineering, telecom applications.

Inet, Inc

(972) 578-6100 (800) WOW-INET

Fax: (972) 578-6113

www.inetinc.com

Inference

(415) 615-7900

Fax: (415) 615-7901

www.inference.com

CBR Express, case-based problem resolution engine for help desks.

Infinet Inc.

(972) 701-9484

www.infinet1.com

Local and Wide Area Networks; CTI; remote office connectivity; network management; sales service support.

Infinite Technologies

(410) 363-1097 (800) 678-1097

Fax: (410) 363-3779

www.ihub.com

E-mail, internet and remote access products.

Info Group

(508) 628-4500 800-54-GROUP

Fax: (508) 628-4566

www.infogrp.com

Telemanagement and call center information systems.

Info Systems

(416) 665-7638 (800) 825-5434

Fax: (416) 665-4193

www.talkie.com

DOS and NT-based voice processing application generator called Talkie-Globe.

InfoActiv, Inc.

(617) 557-4066 (888) 226-6286

www.infoactiv.com

InfoActiv provides consulting and systems integration in call center, computer telephony, interactive voice response, voice messaging, enterprise and operations management.

Infobase Services Inc. (ISI)

(561) 681-7061 (800) 775-5898

Fax: (561) 688-9410

www.ctiguys.com

CTI systems. CTI-Link integration system, Monitoring and routing systems. Systems integration services.

Information Gateways Corp.

(703) 760-0000

Fax: (703) 760-0098

Criterion line of "switchless" call center platforms that incorporate ACD, PBX, IVR, dialing, scripting and campaign management functions.

Information Management Associates (IMA)

(203) 925-6800 (800) 776-0462

Fax: (203) 925-1170

www.imaedge.com

Enterprise customer interaction software (called Edge) for call centers used for sales, marketing and customer service applications. The company also offers a range of professional consulting, technical support, and education services.

Inforte

(312) 540-0900

www.inforte.com

IT/systems integration consultancy.

InfoServ USA

(716) 242-0371

Fax: (716) 242-0617

www.infoservusa.com

Specializes in the design of IVR systems and applications for vertical markets, including automotive ("Auto Credit Hotlines"), finance ("Loan by Phone") and real estate.

Innings Telecom

(800) 363-4223

Fax: (905) 470-8114

www.innings.com

Input

(650) 961-3300

Fax: (650) 961-3966

www.input.com

Intecom

(972) 855-8000 (800) 468-3945

Fax: (972) 855-8533

www.intecom.com

ACD and PBX functions on a single comm platform; the Intecom E is a server-based, standards-based comm platform with unlimited geographical network capabilities with a single database and central administration. it can sustain 3 million CTI messages and more than 500,000 calls an hour. CallWise is the real-time call management system for call routing, CTI, reports, screen sync and call blending. It's modular and upgradable.

IntegreTel

(408) 362-4000

www.integretel.com

Billing and collection services for the telephone industry; call center outsourcing.

Intek Information

(800) 720-5432

Fax: (510) 371-3187

www.intekinfo.com

Custom-designed "high-end" outsourcing services and technology consulting.

Intellisystems

(702) 828-2803 (800) 637-8400

Fax: (702) 828-2828

www.intellisystems.com

Intellisystems markets an interactive expert system that providesself-support on the phone and on the Web. The IntelliSystem guidescallers through a series of diagnostic questions and deliverstrouble-shooting, much like a human expert. Self-support with theIntelliSystem helps callers find their own answers, reducing the numberof issues to be resolved by live agents. This technology is being used successfully to deflect calls from live support at Intuit, Netscape, Gateway 2000, Broderbund, Iomega, and Adaptec.

Interactive Communication Systems

(719) 444-0554

Fax: (719) 444-0150

www.icstelephony.com

Custom services for computer telephony deployment, including things like debit and credit card processing systems, international callback, internet telephony, etc.

Interactive Digital

(516) 724-2323

Fax: (516) 724-2262

www.easytalksoftware.com

EasyTalk add-on software for reducing call duration on Interactive Voice Response Systems. Available for Dialogic/WinNT and MS-DOS Platforms.

Interactive Intelligence

(317) 872-3000

www.inter-intelli.com

Develops a computer telephony product for enterprises and call centers called the Enterprise Interaction Center. EIC is a Java-based communications system bringing for corporate LANs and Intranets. EIC replaces boxes like PBXs, ACDs, VRUs, IVRs, voice mail systems, fax servers, and CTI gateways with a single Windows NT server. Internet integration includes on-line chat, video and VON call back, as well as push technology and automatic Web page generation.

Interactive Quality Services

(612) 820-0778

Fax: (612) 835-2355

www.iq-services.com

Quality assurance consulting and testing services for computer telephone integration and e-commerce systems.

Interalia Communications

(612) 942-6088 (800) 531-0115

Fax: (612) 942-6172

www.interalia-inc.com

Announcement and messaging systems. The XMU Digital Calll Processing system supports up to 63 ports with ACD announcements and automated attendant features. Lots of ways to configure custom announcements through various soft parameters: time of day, holidays, etc.

Interior Concepts

(616) 842-5550 (800) 968-3201

Fax: (616) 846-3985
www.interiorconcepts.com
Furniture systems for call centers; they integrate all the power and cabling connections into their pre-fit systems.

International Customer Service Association
(800) 360-4272
www.icsa.com

International Quality & Productivity Center
(800) 882-8684
www.iqpc.com
Seminars and other educational events.

Interprise
(972) 385-3991
Fax: (972) 788-2441
www.ntrpriz.com
An architectural interior design firm that specializes in call center design and development. Provides planning, design, construction documents, project management, and furniture procurement services. Works nationwide on call center development in new buildings, adaptive reuse or existing facility renovation.

InterVoice
(972) 454-8862
Fax: (972) 454-8905
www.intervoice.com
Advanced call and business process automation systems. OneVoice software agent multi-application platform incorporates auto attendant, IVR. voice messaging, and fax processing. Their systems "allow customers self-serve access to information and transaction processing with database information from a variety of devices including telephones, multimedia PCs, display phones, smart cards, and the Internet."

IntraNext Systems
(888) 638-6398
Fax: (303) 751-5726 www.nextsys.com

Invest In Britain Bureau
(212) 745-0300
Fax: (212) 745-0456
britain-info.org/bis/invest/invest.htm
Location assistance and incentives for American companies looking to place call centers in Britain.

ISC Consultants Inc.
(212) 477-8800
Fax: (212) 477-9895
www.isc.com
Call center consulting and optimization.

ITI Marketing Services
(800) 562-5000 (800) 562-5000
Fax: (402) 392-9423
itimarketing.com
Customer care, service, retaining and growing existing client bases. Also, customer capture, prospecting, selling, and processing sales.

Jabra
(619) 622-0764 (800) 327-2230
Fax: (619) 622-0353
www.jabra.com
Headsets — theirs is a special "in-the-ear" model that combines the speaker and the mike into one tiny piece called the Ear Phone.

Kaset International
(800) 735-2738
Fax: (813) 971-3511
www.kaset.com
Customer service training programs.

Kathy Sisk Enterprises
(209) 323-1472 (800) 477-1278
Fax: (209) 323-9151
www.telenews.com

Key Voice Technologies
(941) 922-3800 (800) 419-3800
Fax: (941) 925-7278
www.keyvoice.com
Small office and corporate voice systems.

King TeleServices
(800) 793-8745
Kirvan and Associates
(609) 228-7525
Disaster recovery consultants and services.

Knowlix Corp.
(800) 784-3060
Fax: (801) 924-6110
www.knowlix.com
Knowlix builds knowledge tools that seamlessly integrate into the existing workflow of internal help desks and customer support centers.

KSBA Architects
(412) 252-1500
Fax: (412) 252-1510
www.ksba.com
A Pittsburgh, PA-based firm specializing in architecture, planning, interior design and project management of call centers with emphasis on "Performance Design," the link between facility design and profitability.

Lernout & Hauspie
(617) 238-0960 888-LERNOUT
Fax: (617) 238-0986
www.lhs.com
Speech recognition, text-to-speech, speech-to-text systems. Featuring speaker independent recognition, barge-in, echo cancellation and line adaptation.

Linkon
(203) 319-3175
Fax: (203) 319-3150
www.linkon.com
Makes a variety of interesting products, including voice boards with a truly unique "universal port" architecture that support a wide range of advanced voice processing apps; and an all-in-one VRU/IVR combo called Escape, that scales well and is a robust, easily upgradable system.

Locate In Scotland Call Center
www.scotcall.com
Assistance in setting up call centers in Scotland.

Locus Dialogue
(514) 954-3804
Fax: (514) 954-3805
www.locus.ca
Speech recognition system incorporated into a "virtual attendant."

Longview Economic Development Corp.
(903) 753-7878 (800) 952-2613

Fax: (903) 753-3646
www.ledco.com

Lucent Technologies
(800) 372-2447
www.lucent.com
AT&T telecom systems spinoff. The inheritor of all that was call center hardware coming out of AT&T: ACDs, voice processing systems, and so on. The home of Conversant and Definity.

Malibu Software
(310) 455-3327
Fax: (310) 456-6225

Manitoba Call Centre Team
(800) 463-6360
Fax: (204) 943-0031
Economic development organization for call center industry in Western Canada.

Manpower
Fax: (414) 906-7837
www.manpower.com
Human resources; training and staffing for call centers (among many other industries they serve).

Maritime Tel & Tel
(902) 421-5884
Fax: (902) 429-4983
Economic development organization for call center industry in Eastern Canada (The Maritime Provinces).

MarkeTel Systems
(800) 289-8616
www.predictivedialers.com
Predictive dialing systems.

MarketFax
(914) 591-6301
Fax: (914) 591-0017

Mastermind Technologies
(703) 276-9300 (800) 644-7605
Fax: (703) 276-9301
www.mastermind-tech.com MasterVox telephony application development platform.

MasterWorks, Inc.
(800) 726-4790
Fax: (781) 246-1328
www.masterworksinc.com

MasterX Corporation
(860) 343-6024
Fax: (860) 343-6028
www.masterx.com
Offers CM-Link and DataMit, two tool suites that facilitate the real-time transport of data from EIS's Call Manager system into Microsoft's SQL Server. (That is, from a Unix environment into Windows NT, where data can be analyzed using a much richer set of tools). Provides lots of benefits, including centralizing data (one point of access for reporting, backup and administration) and associates migrated data with digital voice recordings.

Matrixx Marketing
(513) 397-6864 800-MATRIXX
Fax: (513) 723-9030
www.matrixx.com
MATRIXX Marketing provides outsourced customer management solutions.

Maxxar
(248) 615-1414
Fax: (248) 615-4499
www.maxxar.com
Centrum 9000 platform for running call center and computer telephony apps, like voice response systems, IVR, etc. The system is NT-based. Also makes the MaxxARTS graphical application generator.

MCI Call Center Solutions
(770) 284-4557 (800) 211-8007
Fax: (770) 284-5979
www.mci.com
Offers a full spectrum of services for the call center industry, ranging from the obvious long distance and toll-free services, to intelligent network services and call center consulting. Their own call centers are some of the most advanced in the world.

MCK Communications
(800) 661-2625
Fax: (403) 247-9078
www.mck.com
Remote-voice systems. That is, products that enable remote workers to access corporate

data and voice resources from off-site.

MediaPhonics
(214) 321-5130

Fax: (214) 321-7421

www.mediaphonics.com MediaPhonics designs hardware, firmware and software CTI architectures and products for computer telephony applications.

Mediasoft Telecom
(514) 731-3838

Fax: (514) 731-3833

www.mediasoft.com

IVS Builder and IVS Server are application generators for building IVR apps. The app gen itself is Windows based, and the server runs them on an NT or UNIX platform. Works with most of Dialogic's SCSA product offerings.

Melita International
(770) 239-4000

Fax: (770) 239-4489

www.melita.com

Predictive dialing systems. Melita was one of the first outbound-oriented companies to try to "retro-fit" their image and their products for the increasingly inbound-centric call center world. The major result: call blending, the mixing of inbound and outbound agents, which Melita's PhoneFrame dialer does very well.

Mentor Networks
(800) 585-3129

Mercom
(201) 507-8800

Fax: (201) 507-5554

www.mercom.com

Audiolog voice logging server system.

Meridien Research
(617) 796-2800

Fax: (617) 796-2850

www.meridien-research.com

META Group
(203) 973-6700

Fax: (203) 359-8066

www.metagroup.com
Market research, much of it call center related.

Metasound
(408) 654-0300 (800) 276-2322
Fax: (408) 654-0304
www.metasound.com/
Messaging on hold systems.

Metromail
(708) 574-3800 (800) 927-2238
www.metromail.com
List and lookup services. Acquired by GUS of Great Britain in 1998.

Micro Computer Systems
(972) 659-1514
Fax: (972) 659-1624
www.mcsdallas.com
Makes a product for routing inbound e-mail to reps in support environments. Keeps track of the e-mail and the customer data associated with it.

MicroAge
(800) 246-4322
Fax: (602) 366-1775
www.microage.com
Technology sales and distribution company that runs a lot of call centers.

MicroAutomation
(703) 378-7000
www.microaut.com
Call management software called the CallCenter Millennium, which does screen pop, voice/data transfer, routing, etc. on NT, OS/2 or RS/6000/AIX; also makes a statistical analysis and reporting package for same.

Microlog
(301) 428-9100
Fax: (301) 540-5557
www.mlog.com
IVR systems for the UNIX and DOS worlds.

Mitel
(613) 592-2122

Fax: (613) 592-4784

www.mitel.com

PBXs that have the ability to spin off and run ACD groups for small call centers.

Molloy Group

(973) 540-1212

Fax: (973) 292-9407

www.molloy.com

Knowledge Bridge is an "enterprise knowledge management software solution for customer support". Integrates with client/server call management systems, or standalone problem resolution.

Mosaix

(888) 466-7249

Fax: (206) 558-6001

www.mosaix.com

Predictive dialing systems.

Multi-Channel Systems, Inc.

(800) 724-8340

Fax: (972) 640-0088

www.mcsmk8.com

Makes the MARK VIII PC-based predictive dialer: 8 lines, 5 operators for small telemarketing call centers.

MultiCall

(408) 748-1245

Fax: (408) 748-1257

www.multicall.com

MultiCall makes computer-telephony solutions that address the needs of companies with 5 to 200 agents, whether they are part of a formal call center environment or are virtual agents serving customer needs based on their skill sets.

Music Telecom

(908) 684-1300

Fax: (908) 684-0426

www.music.com

Mustang Software

(661) 873-2500

Fax: (661) 873-2599

www.mustang.com

E-mail management solutions through a combination their Web Essentials tools and its e-mail management tools that include the Internet Message Center, ListCaster and FileCenter.

N-Soft
Fax: (847) 981-5006
www.n-softna.com
Etrog is a family of CTI modules and a CT development environment.

National Consulting Systems
(800) 252-7334
Call center location assistance for many US areas.

National Data Systems
+27 273734324
Fax: +27 273734346
Call centers and cc solutions.

National Service Direct, Inc.
(404) 256-4673 www.nsdi.com

National TechTeam
(800) 522-4451
Fax: (313) 277-6409
www.techteam.com
Computer and customer service solutions including call center services, training, systems integration, and technical staffing. Outsourcing.

Natural MicroSystems
(508) 620-9300 (800) 533-6120
Fax: (508) 650-1351
www.nmss.com
A major supplier of voice boards and voice processing platforms for use in call center and CTI applications.

NBTel/New Brunswick
(506) 694-6022 (800) 824-7449
Fax: (506) 658-7827
www.callcenter.nbtel.nb.ca
The Canadian Province of New Brunswick and that area's telephone service provider, which is trying to attract call centers to come and locate there.

NCR
(937) 445-5000
Fax: (937) 445-5541
www.ncr.com

NEC America
(800) 338-9549
www.nec.com

Net Perceptions
(612) 903-9424 (800) 466-0711
Fax: (612) 903-9425
www.netperceptions.com

Netaccess
(800) 950-4836
www.netacc.com
ISDN and modem technology; cards to fit a variety of platforms.

netDialog
(650) 372-1200
Fax: (650) 372-1201
www.netdialog.com
Web-based customer interaction front end for call centers.

NetPhone
(508) 787-1000
Fax: (508) 787-1030
www.netphone.com
Makes CTI servers, boards and applications for NT.

Netphonic
(415) 962-1111
Fax: (415) 962-1370
www.netphonic.com
They make something called the Web-On-Call voice browser, which claims to integrate
the Internet with an IVR system.

NetSpeak Corporation
(561) 998-8700
Fax: (561) 997-2401
www.netspeak.com

Network Associates

(408) 988-3832

Fax: (408) 970-9727

www.nai.com/

Software company for help desk, data security, etc. Owns Magic Solutions.

Neuron Data

(650) 528-3450 (800) 876-4900

Fax: (650) 943-2752

www.elements.com

Makes a call center automation system that applies business rules to call center processes. Product is called Elements.

NewMetrics Corporation

Fax: (314) 725-0307

www.newmetrics.com

Work force management solutions and services for call center managers.

NICE Systems, Inc.

(201) 617-8800 (888) 577-6423

Fax: (201) 617-9898

www.nice.com

Digital call logging systems that integrate with all major switches and CTI servers, and that scale up to 5,000 channels. They also make a fax management system and unified messaging.

Noble Systems

888-8NOBLE8

Fax: (404) 851-1421

www.noblesys.com

Customized Call Center Automation — that is, a system with inbound, predictive dialing, and blended call management.

Norrell Corporation

(404) 240-3000

Fax: (404) 240-3959

www.norrell.com

Call center outsourcing services. (Under the Norcross Teleservices subsidiary.)

Nortel

(408) 988-5550 (800) 466-7835

Fax: (408) 565-3474

www.nortel.com
Switches, ACDs, software. Also owns Brock Telecom. Latest offering is Symposium, a portfolio of products and services for that provide multimedia call center systems.

North Highland Company
(404) 238-0699
Fax: (404) 233-4930
www.north-highland.com
Management and technology consulting including call center consulting services.

Nuance Communications
(415) 462-8200
Fax: (415) 462-8201
www.nuance.com
Speech recognition system that has a large vocabulary and a pretty good natural language processing system. Schwab has used it for their phone-based stock quote and trading service.

Nuera
(619) 625-2400
Fax: (619) 625-2422
www.nuera.com
Makes digitical circuit multiplication equipment.

Octel
(800) 444-5590
Fax: (408) 324-2632
www.octel.com
Voice systems of all stripes, from VRU to IVR and then some. Owned by Lucent.

Of Eagles and Geese
+61 (39) 6703600
Fax: +61 (39) 6708550
www.oeag.com.au
Consulting firm in Australia that also publishes the The Teleconsultant (a magazine about call centers) and performs training and skills assessments and audits.

Olsten Corp.
(516) 844-7800
Fax: (516) 844-7363
www.olsten.com

Omtool
(603) 898-8900 (800) 886-7845
Fax: (603) 890-6756
www.omtool.com
Makes an Internet/intranet fax server system for enterprise-wide faxing across multiple platforms.

One Call Systems
(800) 845-9945

OnQueue Call Center Consulting
(215) 491-4636
Fax: (215) 491-4636
Provider of workforce management services to call centers.

Ontario Systems
(800) 283-3227
Fax: (317) 751-7198
www.ontario.com
PC-based predictive dialers.

Onyx Software
(425) 451-8060
Fax: (425) 990-3343
www.onyx.com

optimAS AG
Fax: +41 (1) 835 78 79
www.optimas-group.com
Inbound and outbound call center services. Consulting for building European centers.

Oracle
(650) 506-7000 (800) ORACLE1
www.oracle.com
Database software, and more recently, the purchaser of Versatility.

Orion Market Intelligence
+613 9654 4233
Fax: +613 9654 4811
www.omi.com.au
Computer Aided Telephone Interviewer that uses the Microsoft BackOffice pack.

Outreach Technologies
(410) 792-8000
Fax: (410) 792-8008
www.outreachtech.com
Conferencing technology that is an add-on to Microsoft Netmeeting.

Ovum
(781) 272-6414 (800) 642-6886
Fax: (781) 272-7446
www.ovum.com
Consulting and market research about technology.

PakNetX
(603) 890-6616
Fax: (603) 870-9395
www.paknetx.com
Internet-based ACD systems.

Panamax
(415) 499-3900 (800) 472-5555
Fax: (415) 472-5540
www.panamax.com
Surge protectors and line of power related systems for protecting phone systems, networks and PCs.

Para Systems, Inc.
(972) 446-7363 (800) 238-7272
Fax: (972) 446-9011
www.minuteman.com
Power protection devices.

Parity Software Development
(415) 332-5656
Fax: (415) 332-5657
www.paritysw.com
Software tools and hardware components for developing computer telephony applications.

PaylinX Corporation
(314) 692-0929
Fax: (314) 692-0805
www.paylinx.com

Real time credit card authorization software for call center, IVR, Internet, and POS applications.

Peachtree Software
(800) 247-3224
Fax: (770) 564-5888
www.peachtree.com

Pegasystems
(617) 374-9600
Fax: (617) 374-9620
www.pegasystems.com
"Customer interaction solutions": voice and data integration, screen pop, etc.

Pelorus Group
(908) 707-1121
Fax: (908) 707-1135
www.pelorus-group.com/

Perimeter Technology
(603) 645-1616 (800) 645-1650
Fax: (603) 645-1424
www.perimetertechnology.com
A centrex and SL-1 ACD product, plus an ACD Management Information system for those telecom platforms.

Periphonics
(516) 468-0800
Fax: (516) 981-2689
www.peri.com
Voice processing, IVR that also includes transaction processing capabilities and application development tools. And the Internet.

Phonetic Systems
(781) 229-5823
www.phoneticsystems.com

Picazo Communications
(408) 383-9300
Fax: (408) 383-0136
www.picazo.com
PC-based phone system that includes ACD and goes up to 192 ports. Includes various CTI options as well.

Pipkins

(314) 469-6106

Fax: (314) 469-0841

www.pipkins.com

Call center management software, workforce management. Merlang ("modified erlang") is their version of the calculating algorithm. They claim to be able to schedule through skills-based routing, which, if true, is quite a feat.

Plantronics

(408) 426-5858 (800) 544-4660

Fax: (408) 425-8654

www.plantronics.com

Headsets.

Platinum Software Corporation

(800) 883-4582 (800) 883-4582

Fax: (503) 612-2800

www.clientele.com

Platinum Software develops client/server enterprise resource planning software, including customer service, sales automation, financial accounting, budgeting, manufacturing, and distribution for mid-market corporations worldwide. Their Clientele Products Division provides customer service and sales force automation software.

Platinum Technology

(630) 620-5000 (800) 442-6861

Fax: (630) 691-0718

www.platinum.com/

Point Information Systems

(781) 416-2710

Fax: (781) 416-2730

www.pointinfo.com

Portage Communications

(888) 844-5320

Fax: (425) 888-6475

www.portagecommunications.com

Call Center Designer and SimACD, Windows-based software tools for modeling and predicting the operations of an inbound call center. Call Center Designer is for planning the number of people and inbound trunk lines needed for any call volume and desired level of service. You tell Call Center Designer your call volumes, average call lengths, service level goals, and call center costs; Call Center Designer tells you how many staffed agents and trunks you need. Call Center Designer also tells how well your call center is doing and how well it can do by day, hour and minute.

Precision Response Corporation
(800) 666-4772

PriceInteractive
(703) 620-4700 (800) 341-7800
Fax: (703) 758-7108
Offers a third-party verification service for telemarketing applications.

Primus
(206) 292-1000
Fax: (206) 292-1825
www.primus.com
A provider of problem resolution and knowledge management software that enables customer support organizations to capture, share, and manage knowledge. Primary products: SolutionBuilder, a knowledge management and administration app for senior support professionals with solution authoring responsibilities; SolutionExplorer, a Web-based tool for frontline support professionals and remote service providers such as VARs and system integrators; and SolutionPublisher, a Web-based tool that provides customers with self-service support on a corporate Intranet or Internet site.

ProAmerica
(800) 888-9600
Fax: (214) 680-6134
www.proam.com
Service Call Management help desk software.

Product Line, The
(303) 671-8000 (800) 343-4717
Fax: (303) 696-7300
Provides live agent inbound and outbound call handling services, IVR, fulfillment services and transaction processing (credit cards, checks, etc.) to a variety of vertical markets. Primarily involved in customer support and direct marketing applications.

Professional Help Desk
(203) 356-7700 (800) 474-3725
Fax: (203) 356-7900
www.phd.com
Help desk software, featuring call management, call tracking, asset management, problem resolution (via the Help Desk, intranet and Internet), and customized reporting.

Professional Help Desk is Windows-based and fully ODBC compliant, seamlessly integrating with virtually any database and systems architecture.

ProMark One Marketing Services
(800) 933-0233

Promodel Corporation
Fax: (801) 226-6046
www.promodel.com
Simulation product and systems.

Pronexus Inc.
(613) 839-0033
Fax: (613) 839-0035
www.pronexus.com
VBFax fax server, and VBVoice IVR app gen; VBVoice supports up to 192 lines, MVIP and SCSA, and Visual Basic for development. TAPI 2.1 compliant.

ProSci
(970) 203-9332
www.prosci.com

Protocol Communications Services
(800) 564-1444

PTT Telecom Netherlands US
(212) 246-2130
Fax: (212) 246-1905
National telecom carrier of the Netherlands, active in seeking US companies that want to open call centers in that country.

PulsePoint Communications
(805) 566-2255
Fax: (805) 684-2848
www.plpt.com

Purdue University Center for Customer-Driven Q
(765) 494-8357
Fax: (805) 937-4383
www.CallCenterCoach.com

PureSpeech
(617) 441-0000

Fax: (617) 441-0001

www.speech.com

Speech recognition product suite for high volume apps.

Q.Sys

(513) 745-8070

Fax: (513) 745-8077

www.qsys.com

Call Producer telephony server.

Qronus Interactive

(408) 822-5200

Fax: (408) 822-5515

www.qronus.com

Testing systems for CTI products.

Quintus Corporation

(510) 624-2800 (800) 337-8941

Fax: (510) 770-1377

www.quintus.com

Help desk software. Offers a web-enabled product with advanced data "publishing" features in the latest version. Owns Nabnasset & Call Center Enterprises.

Racal Recorders

(703) 709-7114 (800) 553-8279

Fax: (703) 709-9529

www.racalrecord.com

Offers the WordNet voice logger, with up to 96 channel capacity. Available with ISDN PRI or BRI, PCM30 and PCM32 digital interfaces. A Workstation app provides full replay and system administration.

Reese Brothers, Inc.

(800) 365-3500

Registry, Inc., The

(800) 255-9119

www.tri.com

Remedy

(650) 254-4919

Fax: (650) 903-9001

www.remedy.com

Help desk and customer service software.

Renaissance Consulting
(847) 706-6740
Fax: (847) 706-6742
www.consultrenaissance.com
Management consulting firm providing strategic planning and business development services to electronic commerce firms.

Resource Center for Customer Service Professionals
(708) 246-0320
Fax: (708) 246-0251
www.the-resource-center.com
Educational books, videos, seminars, and conferences for call center andcustomer service professionals.

Response Design Corporation
Fax: (609) 399-5311
www.responsedesign.com

Response Interactive
(416) 969-7890
www.responseinc.com
Response has a software product called WebExchange. This provides a live link between visitors to a web site and the appropriate agents from a company's call center. There is a client and server application, as well as a www component. WebExchange uses (optional) a Dialogic board for making audiotext calls, and performing certain telephony functions.

Rhetorex
(408) 370-0881
Fax: (408) 370-1171
www.rhetorex.com
Voice boards, owned by Octel.

Right Now Technologies
(406) 582-9341 (888) 418-3400
Fax: (406) 586-9626
www.rightnowtech.com

RightFax
(520) 320-7000

Fax: (520) 321-7456
www.rightfax.com
Makes a line of enterprise fax servers.

RMH Teleservices
(610) 520-5300 (800) 367-5733
Fax: (610) 520-5356

Rockwell
(630) 227-8000 (800) 416-8199
Fax: (630) 227-8186
www.ec.rockwell.com
Integrated call center platform technologies, including ACD, CTI and information management tools. Galaxy ACD, Spectrum ACD and the Gateway Architecture for Call Centers. Rockwell's ACD systems are good high-volume inbound switches. They are often found in the largest call centers, including the punishingly busy reservation centers.

Ron Weber And Associates
(800) 835-6584

Royalblue Technologies
(212) 269-9000
Fax: (212) 785-4327
www.royalblue.com/

Ruppman Marketing Technologies
(800) 787-7626
Fax: (201) 812-1961

Sage Research
(508) 655-5400
Fax: (508) 655-5516
www.sageresearch.com

ServiceWare
(412) 826-1158 (800) 572-5748
Fax: (412) 826-0577
www.serviceware.com
Knowledge Paks, portable help desk knowledge for adding to a software engine.

Shark Multimedia
(408) 987-5420 (800) 800-3321

Fax: (408) 987-5424

www.sharkrmm.com

Voice messaging and data/fax communications system for "all-in-one" business telephony use.

Siebel Systems

(650) 295-5000

Fax: (650) 295-5111

www.siebel.com

SFA and enterprise-wide customer management systems, and now the owner of Scopus (one of the many help desk-turned-CIS vendors). Includes application software for field sales, customer service, telesales,telemarketing, field service, third-party resellers and Internet based e-commerce and self service.

Siemens

(408) 492-2000 (800) 765-6123

Fax: (408) 777-4988

www.siemenscom.com

Switches, large, high-end ACDs, software for call routing.

Signature Telemarketing

(800) 521-6202

Silknet Software

(603) 625-0070

Fax: (603) 625-0428

www.silknet.com

Silknet's eService 98 is an Internet-based customer service application.

SiteBridge

(212) 645-8700

Fax: (212) 352-2079

www.sitebridge.com

Sitel Corporation

(402) 963-6810 800-25-SITEL

Fax: (402) 963-3004

www.sitel.com

Outsourced telephone based customer service and sales.

Skywave

(408) 245-1771

Fax: (408) 245-1769

www.skywave.net

IP telephony gateway for service providers, as well as a control interface product and a network manager (for things like billing, etc.)

Society of Consumer Affairs Professionals

(703) 519-3700

Fax: (703) 549-4886

www.socap.org

Society of Telecommunications Consultants

(800) 782-7670

Fax: (408) 659-0144

www.stcconsultants.org

Softbase Systems Inc.

(800) 669-7076

www.netlert.com

Makes Netlert, a non-intrusive desktop messaging system for intranets originally developed for call centers.

Software Artistry

(317) 843-1663 (800) 795-1993

Fax: (317) 843-7477

www.softart.com

Help desk system.

Soundlogic

(604) 291-9989

Fax: (604) 291-9949

www.soundlogic.net

Help desk system.

Spanlink

(612) 362-8000 (800) 452-8349

Fax: (612) 362-8300

www.spanlink.com

Internet call center products and services. FastCall Enterprise, a voice processing CTI combo.

Specialized Resources

(972) 664-6600 (888) 774-2400

Fax: (972) 669-2471

www.sritelecom.com
Telecom consulting, systems integration and systems maintenance.

Spectrum Consulting
(210) 349-4380
Fax: (210) 349-4389
www.sgc-inc.com

Spectrum Corp.
(713) 944-6200 (800) 392-5050
Fax: (713) 944-1290
www.specorp.com
Wallboards used to communicate ACD info to agents in a call center. More than 20 models of color and LED wallboards, with real-time stats, color alerts, and user-programmable messages.

SpeechSoft, Inc.
(609) 466-1100 (800) 878-8117
Fax: (609) 466-0757
www.speechsoft.com
Speech Master IVR and app gen system.

SpeechWorks
(617) 428-4444
Fax: (617) 225-0322
www.speechworks.com
Interactive speech systems for automating telephone transactions. SpeechWorks, a software development environment for building advanced telephone-based speech recognition applications. SpeechWorks' provides developers with the ability to "speech-enable" a broad range of transaction processing, information exchange, and messaging applications.

Sphere Communications
(847) 247-8200 (888) SPHEREC
Fax: (847) 247-8290
www.spherecom.com/home.htm

Sprint
(913) 624-3697
Fax: (913) 624-3080
www.sprint.com
Long distance and consulting services for call centers.

SPS Payment Systems
(847) 405-3400
Fax: (847) 405-4856
www.spspay.com
Call center outsourcing that includes a variety of services that centers could use: help desk and technical support, order management (with reps trained in cross-selling and up-selling) and customer service. They handled 65 million customer contacts for clients in the last year.

Square D EPE Technologies
(714) 557-1636
Power protection systems.

St. Lucie Chamber of Commerce
(561) 595-9999 (888) 785-8243
Fax: (561) 335-4446
www.co.st-lucie.fl.us

StarVox, Inc.
(408) 383-9900
www.starvox.com
StarVox's StarGate Server, from StarVox, is a business-to-business VoIP system for enterprise wide deployment. Integrates traditional PBX features with public switched telephone network (PSTN) reliability over a corporate WAN.

State of the Art
Fax: (213) 782-7506
www.hispaniccallcenter.com
Call center outsourcing service bureau specializing in hispanic/Spanish language marketing.

StepUp Software
(214) 352-9424
Fax: (214) 357-3884
www.stepupsoftware.com
Simple helpdesk for small centers.

Steve Sibulsky Productions
(208) 765-4957 (888) 664-6536
Fax: (208) 667-9792
www.onhold6.com/
Message-on-hold production services.

Summa Four
(603) 625-4050 (800) 544-3440
Fax: (603) 668-4491
www.summafour.com
Open programmable switches.

Sun Microsystems
(650) 960-1300
www.sun.com

Sunburst Software
(310) 791-5060
Fax: (310) 791-5062
www.sunburstsoftware.com

SunGard Data Systems
(610) 341-8700
www.sungard.com
Disaster recovery and service assurance programs.

Swisscom
+41-31-34-20202
Fax: +41-31-34-23917
www.swisscom.com
Swiss national telecom carrier, which provides call center solutions (consulting, engineering, support and finance) as well as services and call center products.

Switchview
(800) 845-9989
Fax: (972) 918-9969
www.switchview.com

Sykes Enterprises
(813) 274-1000
Fax: (813) 273-4545
www.sykes.com
Outsourcing company.

Symon Communications
(281) 240-5555 (800) 827-9666
Fax: (281) 240-4895
www.symon.com

Readerboards, and middleware for all sorts of client/server applications that run in call centers, including networking among multi-vendor ACD environments. Interesting stuff.

Syntellect
(770) 587-0700 (800) 347-9907
Fax: (770) 587-0589
www.syntellect.com
Voice processing, IVR, predictive dialing and web-related call center products. They also run a voice and data processing outsourcer (Syntellect Interactive Services) and through the acquisition of Telecorp Systems, they offer Home Ticket Intelligent ANI, a pay-per-view ordering processing service.

Systems Modeling
(412) 741-3727
Fax: (412) 741-5635
www.sm.com
They make Call$im, a simulation tool for modeling call center performance over the medium and long term, helping you figure out what the impact of change will be. Not exactly a workforce management tool, But increasingly simulators are having to stand in for workforce planners because of the complex (and non-random) nature of call delivery systems.

T-Netix
(303) 790-9111 (800) 531-4245
Fax: (303) 790-9540
www.t-netix.com

TAB Products
(650) 852-2400 (800) 676-3109
Fax: (650) 852-2687
www.tabproducts.com
Call center facilities and design systems.

Talx Technologies
(314) 434-0046
Fax: (314) 434-9205
www.talx.com

Tandem Computers
(408) 285-6000 800-NONSTOP
Fax: (408) 285-0505
www.tandem.com

High-reliability servers and platforms for running call center applications. Company is owned by Compaq.

Target Vision
(716) 248-0550
Fax: (716) 248-2354
www.targetvision.com
TargetVision provides employee communication concepts, services and software that allow you to create, manage, and distribute company news in a timely slide show format through TVs, LANs, and intranets.

Taske Technology
(414) 462-0100
Fax: (414) 462-0101
www.taske.com
Taske Toolbox, enhancement for small ACD for reports, management, supervisor screens in real time.

TCIM Services, Inc.
(800) 333-2255

TCS Management Group
(615) 221-6800
Fax: (615) 221-6810
www.tcsmgmt.com
Workforce management system called the TeleCenter System, and a companion NetForce system for multi-site call centers. Forecasts future call volumes, determines arrival patterns, and figures the most efficient scheduling of staff. Also includes modules for Real-Time Schedule Adherence; Skills-based-routing Simulator. Company is owned by Aspect.

Team Alberta
(403) 422-7201 (800) 745-3823
Fax: (403) 427-5924
www.edt.gov.ab.ca
Team Alberta is the agency responsible for marketing Alberta to the call center industry as an attractive location for call centers.

Technology Solutions Company
(312) 228-4500 (800) 759-2250
Fax: (312) 228-4501
www.techsol.com
TSC is an international consulting and systems integration firm. They have relationships

with many key call center suppliers, service providers and analysts.

Teesland Group
+44 (1232) 382122 +(800) 83375263 (in UK)
Fax: +44 (1232) 382118
www.teesland.co.uk
Call Centre property/facilities developer and investor. Teeslands develop flexible, efficient and economical and build-to-suit facilities. Teesland works closely with many government agencies and can assist in procuring valuable packages of selective financial assistance.

Teknekron Infoswitch
(817) 262-3100 800-TEKNEKRON
Fax: (817) 262-3300
www.teknekron.com
Call center software for job applicant screening, agent monitoring and evaluation.

Tekno Industries
(708) 766-6960
Fax: (708) 766-6533
Call center network management system; reports on status, service level.

Tekton
(910) 855-3112 (800) 888-1778
Fax: (910) 855-3220
www.tekton.com
NiceLog digital voice logging systems for recording and archiving telephone transactions.

Tel-Save
Handles the call center needs of the AOL long distance service offering.

TelAthena Systems
(888) 777-7565 x3717
Fax: (212) 206-1963
www.telathena.com

Telecom Ireland
(800) 445-4475
Fax: (203) 363-7176
www.telecomireland.com

Telecorp Products
(810) 960-1000 (800) 634-1012

Fax: (810) 960-1085
www.telecorpproducts.com
AgentWatch, ACD management system with readerboard.

TeleDirect International
(800) 531-6440

Telegenisys
(510) 210-8900
Fax: (510) 256-6660
www.telegenisys.com
Call processing systems that include CTI apps, predictive dialing and IVR.

Telegenix
(800) 424-5220
Fax: (609) 424-0889
www.telegenix.com

Telegra Corp.
(408) 970-9200
Fax: (408) 970-9242
www.telegra.com
Telegra makes the FaxTrace series of fax test equipment — it analyzes Internet fax, fax servers and networks that carry fax traffic. Also available is the Fax Collector, a system for archiving and retrieving all traffic from machines and servers — an excellent idea for fax-based order processing applications.

Telephone Consultancy & Training
Fax: +44 (0) 01483 888 678
www.tctc.co.uk
Consultancy, training, recruitment and performance audit company.

Telephonetics
(305) 625-0332 (800) 446-5366
Fax: (305) 625-3026
www.telephonetics.com
Algorhythms music and message on hold service, with audio production and programming.

Telequest Teleservices
(817) 258-6500 (800) 833-4443
Fax: (817) 258-6505

www.telequest.com
Inbound and outbound telemarketing and teleservices. Call center management, and bilingual capabilities.

Telerx
(215) 347-5700
Fax: (215) 347-5801
www.telerx.com
Teleservice bureau.

TeleSales Inc.
(978) 694-4100
Fax: (978) 694-4850
www.telesalesinc.com

TeleService Resources
(800) 325-2580

Telespectrum Worldwide
(610) 878-7470
www.telespectrum.com
Inbound & outbound telemarketing, customer service, interactive voice response, customer care consulting, call center management, training and consulting services.

TeleTech
(800) 835-3832
Fax: (303) 894-4208
www.teletech.com

Teloquent
(508) 663-7570
Fax: (508) 663-7543
www.teloquent.com
Distributed Call Center (DCC), ISDN-based remote-agent ACD. This product is a technological marvel. It lets you place agents wherever you want — anywhere you can get an ISDN line to them. Small satellite centers, home agents, all become possible with their very clever architecture. Check out the topology diagrams on their home page.

Telrad
800-NEW-PHONE
Fax: (516) 921-8064

Telstra

Fax: (613) 9634 1189

www.telstra.com

Australian telecom carrier.

Teltone Corporation

(425) 487-1515 (800) 426-3926

Fax: (425) 487-2288

www.teltone.com

A variety of telecom/call center products, including interesting telecommuting and agent-at-home systems. Their systems extend the features of digital, proprietary phones (ACD terminals) to remote agents, based upon a "digital softphone", similar to Aspect's Winset or Rockwell's Convergence. They aim for the integrated PBX/ACD market, emulating Lucent, Nortel and Siemens digital phones. All their technology resides at the central-site, within a Windows NT server. All that is needed at the remote site is a TCP/IP-connected PC, and an analog telephone device.

Telus Marketing Services

(888) 867-8498

Fax: (403) 207-2139

www.tms.telus.com

A provider of call center services to businesses, specializing in inbound and outbound sales and customer service marketing programs. They also provide IVR, consulting and market research through The Advisory Group, a division of Telus.

Teubner & Associates

(405) 624-8000

Fax: (405) 624-3010

www.teubner.com

Teubner & Associates makes an advanced fax system called FaxGate, a central fax processing system that is extremely connective and modular. It sits at the intersection of multiple host environments and communication devices — SNA mainframes, for example, or AS/400s, LANs, TCP/IP networks, really as broad a set of hosts as you can ever hope to use. It's a compete system for automating fax applications, and for adding fax to all of a company's back-office transaction processing apps.

Texas Digital Systems

(409) 693-9378

Fax: (409) 764-8650

www.txdigital.com

QuickCom Visual Message Alert System (readerboard). QuickNet software for networking the workstations to the board.

TMSI
(804) 272-6900
Fax: (804) 272-6950
www.tsb.ca
InterLynx CTI suite for call center applications.

TNG TeleSales and Service
(770) 798-9466
Fax: (770) 798-9552
www.worldshowcase.com
Outbound and inbound telephone- and web-based marketing services.

Toshiba America Information Systems
(212) 596-0600
Fax: (714) 583-3499
www.toshiba.com

Trellix Corporation
(781) 788-9400
Fax: (781) 788-9494
www.trellix.com

TriVida Corporation
(310) 736-3700
Fax: (310) 736-3701
www.trivida.com
Data mining products.

TSB International
(416) 622-7010
Fax: (416) 622-3540
www.tsb.ca
PBX data and networking products, including CTI middleware, call accounting and service bureau systems, and PBX data and network management.

Uniden
(817) 858-3300
Fax: (817) 858-3401
Product that interfaces with PBX and desk set telephone to provide wireless headsets.

Unimax Systems
(612) 341-0946 (800) 886-0390

Fax: (612) 338-5436

www.unimax.com

Makes 2nd Nature, a database and "information control" system for PBXs and voice mail systems. Allows you to manage multiple systems from a single access point, and consolidate reporting. Also helps make moves, adds and changes a bit easier.

Unitrac Software

(616) 344-0220

Fax: (616) 344-2027

www.unitrac.com

Unitrac's software applications include: customer service, inbound/outbound telemarketing, campaigns, mass mail management, sales force automation, and fulfillment processing. Scripting is integrated, including inbound scripts that can be used to capture information for call reports, enter data for new prospects and customers, conduct surveys, review accounts, etc.

US West

Fax: (612) 663-3682

www.uswest.com

USA 800

(800) 821-7539

Vanguard Communications

(973) 605-8000

Fax: (973) 605-8329

www.vanguard.net

Vantive

(408) 982-5700

Fax: (408) 982-5710

www.vantive.com

Help desk software, now known more broadly as "front-office automation software" — functions more like an integrated customer management system than anything else.

Venture Development Corporation

(508) 653-9000

Fax: (508) 653-9836

www.vdc-corp.com

Call center market research and consulting.

Venturian Software

(612) 931-2450

Fax: (612) 931-2459
www.venturian.com
IVR; CyberCall web/call center integration system.

Viking Electronics
(715) 386-8861
Fax: (715) 386-4349
www.vikingelectronics.com
Fax/data switches, auto attendants, digital announcers, toll restrictors, auto dialers.

VIP Calling
(781) 229-0011
Fax: (781) 229-7701
www.vipcalling.com
VIP Calling is a carrier that uses the internet to provide wholesale international telecom services.

Virginia Southwest Promise
(540) 889-0381 (800) 735-9999
Fax: (540) 889-1830
www.vaswpromise.com

Virtual Corporation
(973) 927-5454 (800) 944-8478
www.virtual-corp.net

Virtual Hold Technology
(330) 666-1181
Fax: (330) 666-3869
www.virtualhold.com

Vision Quebec
Fax: (888) 288-6884
www.visionquebec.com

Visual Electronics
(303) 639-8450 (800) 622-9893
Fax: (303) 639-8416
www.digital-fax.com
Readerboard systems.

VocalTec Communications
(201) 768-9400

Fax: (201) 768-8893
www.vocaltec.com
IP telephony systems and gateways.

Vodavi
(602) 443-6000
Fax: (602) 483-0144
www.vodavi.com

Voice Control Systems
(972) 726-1200
Fax: (972) 726-1267
www.voicecontrol.com
Speech recognition toolkit.

Voice Processing Corporation
(617) 494-0100
Fax: (617) 494-4970
www.vpro.com
Speech recognition.

Voice Technologies Group
(716) 689-6700
www.vtg.com
PBX integration products, unified messaging and other CTI systems.

Voice Technology Ltd
+64 (9) 356 5555
Fax: +64 (9) 356 5550
www.voicetechnology.com
Call control/CTI/IVR systems for call centers. A New Zealand-based company.

Voiceware Systems
(561) 655-1770
Fax: (561) 655-2104
www.voiceware.net
Call processing systems integrator.

Voysys
(510) 252-1100 800-7-VOYSYS
Fax: (510) 252-1101
www.voysys.com

Computer telephony products for small centers, especially IVR.

VXI
(603) 742-2888 (800) 742-8588
Fax: (603) 742-5065
www.vxicorp.com
Manufactures headsets and microphones for computer and telephony applications.

WebLine Communications
(781) 272-9979
Fax: (781) 272-9989
www.webline.com
A Web-based call center adjunct that combines telephone connectivity with the Web. Core is a Java-based client/server app that lets agents deliver visual info to the customer's Java-enabled browser.

Wesson, Taylor, Wells & Associates
(800) 833-2894
Fax: (919) 941-0082

West TeleServices Corporation
(402) 571-7700 (800) 542-1000
Fax: (402) 571-7000

Western Direct
(800) 400-4042

WildFile Inc.
(612) 551-0881
Fax: (612) 551-9998
www.wildfile.com

Witness Systems
(770) 754-1900
Fax: (770) 754-1888
www.witsys.com
Witness Systems is a developer and supplier of client/server quality monitoring software for call centers. The company's Witness software is deployed by call center managers to enhance enterprise quality, agent efficiency and supervisor productivity.

Wygant Scientific
(503) 227-6901 (800) 688-6423

Fax: (503) 227-8501

www.wygant.com

Wygant makes SmartAnswer, a voice processing application that informs callers of their position in queue and how long they can expect to wait. It also is a building block for other applications like intelligent call routing, data-follow-me, Vision voice mail and IVR. Encore is an NT-based digital call logger for voice and data recording.

Xircom

(805) 376-9300 (800) 438-4526

Fax: (805) 376-9311

www.xircom.com

Networking products.

XL Associates

(301) 770-0090

Fax: (301) 770-2354

www.xla.com

Call center consulting services.

xPect Technologies

(303) 604-1258

Fax: (303) 665-3884

www.xpecttech.com

Provides processes and systems to help maximize the performance of employees in call centers.

Xtend

(212) 951-7600 (800) 231-2556

Fax: (212) 951-7683

www.xtend.com/

Software developer of PC Based CTI and telemanagement systems.

Zacson

(800) 478-6584

www.syn.net/telemkt/zacson/

Ziehl Associates

(516) 437-1300 (800) 654-1066

Fax: (516) 437-1305

www.ziehl.com

Headsets and other small scale telecom equipment.

the call center
DICTIONARY

350 pages • Code 3288 • $19.95

Over 1,200 Comprehensive and Easy-To-Understand Entries Covering Help Desks, Telemarketing, Customer Service, Voice Processing, Switches, Software, Training, and Call Center Management. This book is the key to understanding how call centers work. The language of the call center comes from many fields. It includes terms from technical fields such as telecommunications, engineering, and computer programming, B marking, customer service and quality assurance terms loom large.

The Call Center Dictionary explains them all in a way that is simple enough for a newcomer to grasp, but with enough depth to give insight to an old pro. It's a guide for everyone whose company does business by telephone. It not only tells you what a particular technology is, it tells you how using that technology can improve your relationship with your customers. (Or it warns you about bad techniques that can alienate customers.) It's for every call center manager or supervisor who doesn't always understand the techies' jargon.

1-800-LIBRARY
408 848 3854 • Fax 408 848 5784 •
6600 Silacci • Gilroy, CA 95020

Milton Keynes UK
Ingram Content Group UK Ltd.
UKHW050259161024
449569UK00042B/1815